Introduction to Wildlife Conservation in Farming

Introduction to Wildlife Conservation in Farming

Stephen Burchett

and

Sarah Burchett

A John Wiley & Sons, Ltd., Publication

This edition first published 2011 © 2011 by John Wiley & Sons, Ltd.

Wiley-Blackwell is an imprint of John Wiley & Sons, formed by the merger of Wiley's global Scientific, Technical and Medical business with Blackwell Publishing.

Registered office: John Wiley & Sons, Ltd, The Atrium, Southern Gate, Chichester, West Sussex, PO19 8SQ, UK

Editorial Offices: 9600 Garsington Road, Oxford, OX4 2DQ, UK
The Atrium, Southern Gate, Chichester, West Sussex, PO 19 8SQ, UK

111 River Street, Hoboken, NJ 07030-5774, USA

For details of our global editorial offices, for customer services and for information about how to apply for permission to reuse the copyright material in this book please see our website at www.wiley.com/wiley-blackwell.

Library of Congress Cataloguing-in-Publication Data
Burchett, Stephen.
  Introduction to wildlife conservation in farming / Stephen Burchett and Sarah Burchett.
    p. cm.
  Includes bibliographical references and index.
  ISBN 978-0-470-69935-5 (cloth)/978-0-470-69934-8(paper)
  1. Agricultural conservation. 2. Wildlife conservation. 3. Land use, Rural – Environmental aspects. 4. Agricultural ecology. I. Burchett, Sarah. II. Title.
  S604.5.B87 2011
  639.9 – dc22

2010020658

A catalogue record for this book is available from the British Library.

This book is published in the following electronic formats: ePDF 9780470670101 Wiley Online Library 9780470670095

Set in 11/13pt Minion by Laserwords Private Limited, Chennai, India
Printed and bound in Singapore by Markono Print Media Pte Ltd

First Impression   2011

Contents

Preface

At the beginning of the 21st C the human population stands at around six billion people and is currently rising at an alarming rate in many regions such as Africa and South East Asia. Globally, the number of human beings is expected to reach nine billion by 2050. This continued increase in human population necessitates the expansion of agricultural systems and agricultural technology to ensure food production can meet future demands. Such increases will inevitably be in direct conflict with natural ecosystems, for instance tropical rainforests, and consequently with wildlife conservation. Indeed across the globe people are becoming increasingly aware of the need to protect the environment from the excesses of commercial agriculture, but it is often difficult for emerging economies to align food production systems with conservation efforts as these nations strive to secure their food supplies and to compete in the fiscal economy of the 'global village'.

As mankind's activities continue to shrink and degrade natural habitats, politicians and policy makers have realised that the farmed environment forms an integral component in the conservation of biodiversity. Consequently these parties have started to redefine agricultural policy with a view to mitigating the excess of agricultural pollution and improve the farmed environment for numerous species. Indeed, European Union (EU) politicians have now started to rewrite the Common Agricultural Policy (CAP) following the 1992 Earth Summit in Rio de Janeiro in order to meet Europe's obligations on the conservation of biodiversity. The simple fact is that farming practice encompasses 47% of the European landscape and in the UK this figure is closer to 70%. These early developments in EU policy have made a start at realigning farming practice and environmental protection and has achieved some significant gains in the protection of watercourses, farmland habitats and associated endangered species, one good example is the increase in the number of cirl buntings in South Devon. However, this is a long and rocky road and not all of these developments in policy have resulted in improvements to biodiversity, in particular some arable reversion projects have resulted in losses to rare and vulnerable arable flora. These losses are often the consequence of rigid application of poor policy decisions that fail to recognise the historical co-evolution of species with traditional farming practices. The de-stocking of upland areas of the UK is one such example. Ecologically problems arose from overstocking due to headage payments paid to farmers during the 1970s and 1980s, these payments encouraged such overstocking; however the current policy of de-stocking has gone too far and is now having a direct, negative, impact on the composition of upland flora, as invasive species such as bracken and gorse out-compete smaller and diminutive native upland vegetation. It should not be forgotten that the UK upland habitats that exist today are actually the result of centuries of traditional farming practices.

Further afield governments are starting to follow Europe's lead and there are some positive developments aimed at protecting the environment and the conservation of biodiversity. However, developing nations are expanding crop systems across virgin forests at frightening rates and consequently contemporary developments in policy are often too little and too late as well as being very difficult to enforce. Politicians, ecologists, agricultural scientists and farmers have to continue to work together to improve food production systems to ensure food security for the future, and to protect the environment at the same time.

This volume aims to introduce the concept of wildlife conservation within the farmed environment, and is structured into six chapters that explore key issues and present real life case studies of farmers, ecologists and researchers working together to improve farming practice and conservation of biodiversity. Chapter one sets the scope of the topic while chapters two, three, four and five review the key farming systems (arable, grassland, forestry and the aquatic environment, respectively). Each chapter gives an overview of these topics as well as presenting a series of case studies, that are complemented with maps, species boxes and tables of organisms protected by the application of good farming practice. Chapter six looks to the future and reflects on developing trends in farming practice and technology, and then evaluates some of the shortfalls in education. What becomes apparent is that at the primary school age groups, children are actively encouraged to look at the countryside and engage with wildlife, but that this all comes to an abrupt halt at secondary level. That is, that the development of conservation science lacks rigour for students in the 11 to 18 year old cohort, an issue that has to be addressed at a fundamental level as students enter higher education wishing to study environmental and conservation sciences.

Applying conservation methodologies within farming practice has a long way to go. However the authors have highlighted, using real examples, many ways in which farmers, with the support of politicians and scientists, can make a positive contribution to wildlife conservation at best, and mitigate environmental damage at the very least. This book is aimed at undergraduate students, conservation groups, policy makers and conscientious farmers in the hope of inspiring future practitioners of:

WILDLIFE CONSERVATION IN FARMING

Stephen Burchett
Sarah Burchett

Acknowledgements

The authors would like to acknowledge the following people who have helped us in our research and preparation of the manuscript.

From the UK: Sandra Hughes for her artwork, Jill Sutcliffe of the Institute of Ecology and Environmental Management (IEEM) for advice on manuscript preparation.

UK academic and farming contacts: Richard and Judy Foss of Down Farm, South Devon. Alison and Ian Samuel for discussions on organic farming in South Devon. Sue Peach of Drywall Farm, Widecombe, Devon. Jim Rudderham and Lindsay Hargreaves from Elveden Estate. Jamie Blackett from Arbigland Estate in Dumfries and Galloway. Henry Edmunds of the Cholderton Estate, Hampshire. David Robertson from the Forestry Commission for an informative visit around the Sunart Oakwoods. Mark Woods of Loch Duart Salmon Farm, Scotland. Tony Smith of Dragon Feeds, Wales.

From the US: Michael Wisniewski from the United States Department of Agriculture (USDA) for help and advice on farming in West Virginia and manuscript preparation. Julian Hoekstra and Jeff Adolphsen for informative discussions on the US Easement Programme.

US academic and farming contacts: Kevin Williams of the USDA Extension Service in North Carolina for organising informative visits with long leaf pine forest owners. Art Williams of Kalawi Farm, North Carolina. Brent Bogue (USDA), John Ann Shearer of the Fish and Wildlife Service (FWS) and land owner Jutta Kuenzler for an excellent wetlands restoration visit in North Carolina, plus Matt Kinane, Gerry Cohn of the American Farmland Trust and farmer Larry Perry in North Carolina, farmers Michael and Chrissie James in Maryland. Malin Clyde and Matt Tarr from University of New Hampshire Extension Service, who arranged many interesting and informative visits for us in New Hampshire, and Richard Langan for taking us out to see offshore aquaculture operations. Ned Therrien for showing us around his woodland lot in New Hampshire and farmer Chuck Cox also in New Hampshire. Eric Sideman from the Maine Organic Farmers & Gardeners Association (MOFGA).

From the rest of the world: Glen Reynolds from Danum Valley, Sabah, Borneo. Bill Spencer and Paul Troy of Hawaii Oceanic Technology™. Dr Shamsudin, Forestry and Environment Division of Forestry Research Institute Malaysia, for in-depth discussion on Malaysian Forestry and Enrichment Planting. Kristoffer Hylander from Department of Botany, Stockholm University for his insights on shade coffee growing.

This list is by no means exhaustive and we would also like to thank all of those people who have talked to us about farming and wildlife during our research period, and those who have sent e-mail links introducing further contacts.

1 Introduction

1.1 Conservation on farmland – why?

The population of planet earth stands in excess of 6.6 billion people, more than double what it was 50 years ago, and it is predicted to increase to approximately 9 billion in the next 50 years. That's a lot of mouths to feed, and more of the land will inevitably need to be turned over to farming to supply that demand. As the amount of land used for production increases, so does the impetus for integrating conservation into farming, we cannot think just in terms of designating specific conservation areas and national parks. Such areas are extremely important, but they are simply not enough if we are to get the balance between that which we take out of the ecosphere, and that, which needs to be replenished and replaced, back on track.

There is considerable disparity in the percentage of land each country currently uses for farming. For example, in the United Kingdom (UK) around 23% of land is used for growing arable crops, in India the figure is nearly 50%, but in Egypt it accounts for less than 3%. The reasons are clear when one considers such issues as terrain, sub-strata, the underlying geology that predetermines the bedrock and subsequently the composition of the top soil, population dynamics and urbanisation, however food security has to be addressed on a global scale despite such statistics. In countries where the land is difficult to cultivate, conservation may not come very high up on the list of priorities when considering the need to feed an increasing population – but perhaps it should? Conservation is not just about protecting individual species; it comes as a whole package, where sustainable land management will endeavour to enhance habitat and species conservation while simultaneously producing nutritious and healthy food. Below are a few examples of the, often devastating, consequences when the proper connection between farming and habitat conservation is not made.

- El Salvador, January 2001 and Argentina, February 2009 provide just two of numerous examples where heavy rainfall has created massive mudslides that have killed many people and devastated homes and arable land. Deforestation has been postulated as one possible cause; certainly it will have exacerbated the problem. Soil is bound together by roots such that removing trees can cause significant loss of topsoil which results in both poor soil structure for growing crops, damaging

Introduction to Wildlife Conservation in Farming   Edited by Stephen Burchett and Sarah Burchett
© 2011 John Wiley & Sons, Ltd

run-off into watercourses, (discussed in greater depth in Chapter 5), and, when there is high rainfall, these massive mudslides may result. So could these terrible tragedies have been avoided if large-scale deforestation had not taken place? They certainly believe so in Kyrgyzstan, Central Asia. With the support of Switzerland, large numbers of walnut trees (*Juglans spp.*) have been planted, and existing trees preserved. This has resulted in fewer mudslides and provided a new and sustainable source of income for the local rural populace, (www.sdc.admin.ch – accessed 12 January 2010).

- In the UK, grants were given to farmers to remove hedgerows (Agricultural Act 1947) to increase the area of production and increase the speed of harvesting, larger, more efficient machinery were then able to be used in the larger fields. Approximately $1/4$ million miles of hedgerows were removed as a result. The long term effects of this decision were to increase soil erosion → which thus increased the need for chemical nutrient inputs → thus increasing costs → and so increasing the run-off into water courses → thus damaging streams, rivers and often estuaries, and increasing chemicals in the water table from where our drinking water originates. This is all *before* we consider the ecosystems damaged more directly by the removal of hedges. Hedges support a vast number of species of plants and invertebrates. Many of these invertebrates feed on crop pests, so an important ally was being removed along with the hedgerow plants. In turn, these invertebrates support numerous songbirds, many of which have now become endangered through both the loss of habitat and the increase in inputs mentioned. A whole cycle of damage followed as a result of this dubious decision, and no real long-term benefits resulted. Now in 2010 grants are being awarded to farmers to replant hedgerows.

- Our use of ever increasingly bigger, more high-tech and expensive machinery is so often ludicrously disproportionate to the landscape that it is bludgeoning across. As this equipment becomes more and more expensive to buy, smaller farmers often get together in co-operatives to share such equipment to reduce costs, or perhaps more commonly, hire contractors to carry out the work for them. Over the years this machinery 'overkill' has sometimes resulted in the loss of some of the smaller marginal habitats such as vernal pools, and the specialist plant and animal species that are dependent on such habitats have come under threat. One such rare plant in the UK is the Starfruit plant (*Damasonium alisma*) (Species Box 1.1). Its demise – from 50 ponds in the Home Counties (Fisher, 1991) to some 15 ponds currently in Buckinghamshire and Surrey (www.plantlife.org.uk – accessed 6 March 2010), occurred as a result of farm animals making less use of ponds. The trampling by cattle while drinking enabled the plant to germinate. Now Plantlife volunteers walk into the ponds to replicate the former activity of the animals.

- Overseas, mangrove swamps provide another example. Large areas of mangroves have been felled, mainly for agriculture on the landward side and for shrimp farming on the seaward side, but mangrove systems support a vast array of wildlife.

Species Box 1.1: Starfruit Plant (*Damasonium alisma*)

Profile:

A small plant of water meadows in south-east England, parts of northern Europe, Spain, Asia Minor and northern Africa. It has three white petals, yellow at the base and is defined by the distinctive six-star shaped fruits. This plant has extremely distinct requirements to survive. It will only germinate underwater in vernal ponds; it then puts out fairly long floating leaves. In order to flower the pond must dry out early in the season, to prevent competition, at which time the floating leaves die back and the plant puts out new, more rigid leaves and stems bearing the flowers which are insect pollinated. Seeds may lay dormant for many years until conditions are right. This is a plant that has evolved with farming. The seeds require disturbance to break dormancy that historically occurred when livestock went to drink at the ponds. Paradoxically modern farming practice largely excludes this likelihood now as farmers are discouraged from allowing their stock to drink at the delicate poolsides in lieu of the damage this causes to this environment.

Photo: Starfruit Plant. © Natural England/Peter Wakely

Conservation Status:

- Biodiversity Action Plan (BAP) Species Red List – Endangered.
- Schedule 8 Wildlife & Countryside Act 1981.

Current Status:

Always a rarity in the UK, the starfruit plant declined rapidly over the last few decades, due to habitat loss through changes in farming practice and through competition, until at one point it was recorded in only one pond. It has since been subject to a number of conservation programmes, including Kew's Millennium Seed Bank Project, BAP recovery programme and work by English Nature and Plantlife. Numbers are reported to be growing; though it's highly specific requirements make conservation in the wild potentially tenuous.

References

Fisher, J. (1991) *A Colour Guide to Rare Wildflowers*, Constable, London
www.kew.org
www.english-nature.org.uk
www.plantlife.org.uk
www.ukbap.org.uk

They provide a buffer between land and sea. On the seaward side, they form an important nutrient cycle with coral reefs (Hogarth, 1999), and also sometimes with seagrass (*Thalassia spp.*) beds which are often intermediary between the two. Also the shallow waters in which the mangrove trees grow provide a nursery for juvenile reef fish and invertebrates living in the sea grass beds and on the reef, a place where most predators are just too big to go. The mangroves also help to bind the soil, reducing erosion and provide protection from floods. Where mangroves have been removed, tropical flash floods have swept soil down rivers and out of estuaries smothering and choking both sea grass beds and coral reefs often beyond redemption. On the landward side, mangrove based habitats, the 'Mangal', support numerous epiphytic plants, invertebrates and vertebrate species, but mangrove leaves are low in nutrients and not very digestible so that species living in this community are highly adaptive and for some, mangroves are therefore their primary habitat. From a human perspective, the mangroves' capacity to reduce soil erosion is important for the local farming populations that are growing crops nearby. Also, use of the wood from mangroves is important for local people. Such use can be potentially devastating to the mangrove habitat but harvesting can be carried out sustainably, for example in the Matang Forest of western Peninsula Malaysia the wood is harvested by block rotation and used for making charcoal plus some replanting helps to sustain the cycle (Hogarth, 1999).

Other examples include the Lowland Native Grasslands of Australia, discussed in Chapter 3 and the extensive logging of tropical rainforests in South East Asia discussed in Chapter 4.

1.2   Historical relevance of on-farm conservation

'Old-World' countries, such as Britain and other European nations have farmed the land for over 9000 years. Here species of plants, fungi, microbes and animals have evolved in conjunction with farming practices, examples like the common poppy (*Papaver rhoeas*) and the corn marigold (*Chrysanthemum segetum*), highlighted in Species Box 2.2, are known as arable flora and were brought in from the Mediterranean among the crop seeds. Songbirds such as the cirl bunting (*Emberiza cirlus*), highlighted in Species Box 2.1, have co-evolved with arable crops. Another example is the co-evolution of waterfowl and wet grassland pasture. Here birds of international importance (The Wet Grassland Guide, 1997) like the Bewick swan (*Cygnus columbianus*) the bean goose (*Anser fabalis*) and the wigeon (*Anas penelope*) are partially dependent on wet grassland for overwintering on flooded meadows. The relevance of on-farm conservation is therefore paramount – to exclude species that have become part of the farmland ecosystem is to exclude what has become part of a semi-natural cycle.

This begs the question, is on-farm conservation relevant to the 'New-World'? During our travels researching for this book, we visited New Zealand. From a farming point of view, New Zealand is one of the newest countries in the world, most having only been farmed for less than 200 years even in the oldest settlements, and for less than 50 years in many parts. Clearly native wild species cannot have evolved within these farming systems, indeed many such native species struggle for survival at all amidst the introduced species brought by European settlers.

The New Zealand Government has set aside vast tracts of land as conservation areas, with the aim of protecting native and indigenous species from non-native species in particular predatory species that predate the eggs of flightless birds such as the kiwi (*Apteryx spp*). The settling of the country was accompanied by the introduction of many species not native to the country. The consequence is that in many places conservationists work hard to eradicate non-indigenous species; indeed many of the numerous islands have proven to be ideal for this type of conservation programme. On most areas of the mainland however, they seem to be fighting a losing battle. Driving south on state highway 6 towards Queenstown in late December, flanked by snow-capped mountains, lakes and creeks, the road is lined for mile upon mile by colourful flowers; viper's bugloss (*Echium vulgare*) (Figure 1.1), *Lupinus spp.*, dog roses (*Rosa canina*) and *Verbascum spp.*, fill the roadside and foothills, a stunning sight to behold – all European invaders – and certainly there to stay! However, many areas of natural bush do exist and support indigenous insects and birds. Indeed, as you drive down the road you can have the schizophrenic experience of one side of the road being covered in familiar European species while the opposite side of the road is like nothing you've seen before, including tree ferns and a range of trees. Farmed areas have had the natural bush grubbed up and these areas have been seeded

Figure 1.1 Viper's Bugloss are one of many European plant species seen *en masse* across many parts of New Zealand. (Photo: Sarah Burchett)

with exotic grasses for grazing by the 45 million sheep and 10 million beef and dairy cattle – also introduced – as New Zealand has no indigenous land mammals. Clearly New Zealanders have to eat and just as clearly many native species are largely excluded from farmland.

Thus we return to the question 'Is on-farm conservation relevant to the New World'? Despite the apparent arguments against it, we believe that 'yes' it still is. Most farmland in New Zealand will never again be populated by predominantly indigenous plants and animals, indeed this farmland now supports numerous European songbirds also introduced by settlers who wanted to 'feel at home' in their new country and these birds have evolved historically within farmland communities back in their original homeland, plus there are many native birds and insects are opportunists and benefit from the farmland too.

Conservation of water courses on farmland is recognised to be important as native aquatic species can still be found in abundance in the creeks and rivers criss-crossing the farmers' land. Run-off from farmland may potentially cause harm to these species, and small grants have been awarded to some farmers to plant riparian strips. Legislation also exists, compelling farmers to build bridges or to develop alternative routes for stock movements to passage for stock across watercourses. Unfortunately, though such legislation exists, and some small grants may be awarded, there is at present very little financial incentive for New Zealand farmers to apply conservation measures on their land.

In America, also a 'New World' country farming practices are somewhat different from New Zealand. Though large-scale farming in America is relatively recent compared to Europe and Asia, the first European settlers learned to grow maize (*Zea mays*), a crop indigenous to Mexico, from the Native Americans. Since then many of the major crop species grown in the Americas, such as the blueberry (*Vaccinium spp.*), the sunflower (*Helianthus annuus*) and cotton (*Gossypium spp.*) are native plants that have been cultivated as opposed to introduced species and many of the grazing animals also feed on native grasses. Here then farming and conservation of native wildlife can go comfortably hand in hand given the opportunity to do so.

1.3   Legislation and policy

Making changes to farming practice is a big undertaking. There may be considerable costs involved, not to mention time, available labour and materials and, of course, the will to make such changes. Many farmers recognise the need to conserve the land that they are farming, but do not have the resources or manpower to implement such radical steps. Other farmers may require a little 'educational nudging' to urge them to actually see the point in committing time and effort on such an undertaking.

A further issue, which compounds the problem with respect to wildlife conservation on farmed lands, is that farmers produce their products in response to national and international policy. Since the 1940s agricultural policy across Europe and the USA has been biased towards increasing food production and improving the efficiency of agricultural systems. This is quite logical, as nation states need to secure food supplies for their populations. However during the 1980s food surpluses became a problem in Europe and it was this very fact that instigated change in policy drivers to control food production and maintain commodity prices. Early changes in policy instruments were aimed at reducing the area of certain crops, notably cereals, and impose a quota system on some products like milk. In 1988 the European Union (EU) implemented a voluntary grant aid programme where farmers could set aside 20% of their productive land for five years. In America a similar land diversion scheme was introduced during the 1950s in an attempt to reduce food surpluses (Warren *et al.*, 2008). The subsequent benefits to wildlife conservation were noted during these early attempts at land diversification (set-aside). Set aside provides nesting habitats for partridges (*Perdix perdix*) yellowhammers (*Emberiza citrinella*) reed buntings (*E schoeniclus*) and a refugia for ground beetles and many other invertebrate species (www.rspb.org.uk – accessed 6 March 2010). Land diversification schemes eventually became enshrined in the European Common Agricultural Policy (CAP). However, they had been welcome side effects rather than deliberate goals.

1.4   Impact of agricultural policy on the environment

In 1957 Belgium, France, Germany, Italy, Luxemburg and the Netherlands signed the Treaty of Rome and established the Common Market and from this point forward agriculture in these six member states was strongly affected by state intervention through the CAP. Article 33 (39) enshrines the legal basis of CAP and established a set of internal objectives:

- increase agricultural productivity by promoting technology and optimising labour;
- ensure fair standard of living for farmers;
- stabilise markets;
- assure the availability of supplies;
- ensure reasonable price for consumers.

(www.europarl.europa.eu/factsheet – accessed 7 January 2010)

During the 1950s, and as a consequence of the Second World War, other European countries, such as the UK, were developing their own agricultural policies and infrastructure. The 1947 UK Agricultural Act guaranteed prices for products like cereals, sugar beet, beef and milk; such price guarantees encouraged farmers to expand and modernise their production systems. In 1952, the Agriculture Act paid grants to farmers of £30 ha^{-1} to plough up 12-year-old pasture maintained by grazing animals and convert it to cereal production (Warren *et al.*, 2008). This was a further conversion of permanent pasture in an attempt to increase food production, following an initial conversion of 2 million ha of permanent pasture between 1939 and 1945, as a consequence of the cultivation orders. In the 1970s and 1980s, Britain and many other European countries joined the EU and were subsequently directly affected by CAP.

The impact of CAP and other policy drivers on agriculture was profound and from the late 1950s through to the early 1990s advances in agricultural technology changed the face of farming across Europe to such an extent that extensive traditional mixed farming systems evolved into specialised large scale industrialised farms. The impact of industrialised agriculture on wildlife was overwhelming as large areas of land were subjected to nutrient enrichment from continued inputs of artificial fertiliser (Figure 1.2); increased pesticide use and activities that modified the landscape. These activities included the removal of hedgerows, the conversion of woodlands and heathlands to arable land and intensively managed dairy pasture together with the drainage of wet meadows and flood plains to improve field systems and make them suitable for intense arable crop production. Such changes in land use removed valuable habitats for numerous species which had coevolved with agriculture in the preceding 10 000 years (Warren *et al.*, 2008). These species have now become synonymous with modern wildlife conservation across Europe. Examples include the annual flora of cereal fields (Wilson and King, 2003) such as the catchflies (*Silene noctiflora* and *S gallica*, Species Box 2.3), which have declined in their range and abundance since the advent of modern farming practice. Other examples of co-evolved organisms that have declined in recent times are the farmland birds and in particular the small seed eating birds of the arable landscape, such as the cirl bunting (*Emberiza cirlus*, Species Box 2.1). The list for Europe and the UK is almost endless with examples such as the black grouse (*Tetrao tetrix*), capercaillie (*T urogallus*), red grouse (*Lagopus lagopus*) and the ptarmigan (*L mutus*) in Scotland as land use change was driven by policy

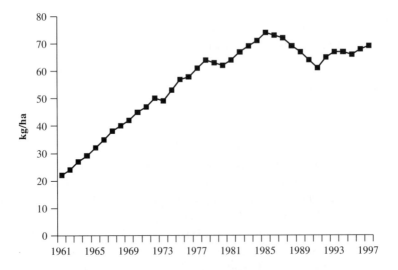

Figure 1.2   Average nitrogen use in the EU from 1961 to 1997

to convert moorlands into commercial softwood plantations and the subsequent application of poor management and pest control practice (Rose, 2004).

In the USA, similar examples of habitat loss and decline in associated species can be observed such as the decline in longleaf pine habitat and associated fauna, in particular the red cockaded woodpecker (*Picoides boralis*, Species Box 4.3). This bird is a specialist species of a natural habitat but it has been highly modified by modern logging practice. Sadly, as the population of humanity increases and the demand for increased standards of living rises, the ecological impact of humanity on natural ecosystems can be devastating. Many highly charismatic and keynote species are in decline, examples include the orang utan (*Pongo pygmaeus*) of Borneo as their forest home is being logged and/or converted to oil palm (*Elaeis guineensis*) at an alarming rate. All the above examples of land use change and associated decline in species are a direct response of landowners to policy edicts from governments aimed at improving agricultural performance and standards of living reflecting a lack of understanding of the importance of the natural world.

In recent years there has been a change in political philosophy across Europe and the USA towards agriculture and wildlife conservation, which is now reflected in policy decision-making processes. In Europe, the all-encompassing CAP has undergone a number of modifications (Table 1.1). There has been an increasing emphasis on the conservation of biodiversity and rural communities, and similar changes have occurred in USA policy (Table 1.2). In the USA the application of rural policy is focused on government renting land from the farmer in an attempt to take land out of cultivation and thus protect fragile habitats. Indeed the approach in the USA reflects a philosophical difference between European and US agriculture. In Europe, agriculture and wildlife conservation are perceived by policy makers as multifunctional uses whereas the US view agriculture and conservation as mutually exclusive (Warren *et al.*, 2008) and this difference is clearly reflected in the approach to on-farm conservation between the two continents.

In a multifunctional model, agriculture is not only responsible for producing food and fibre crops with a monetary value but also ecosystem services such as the maintenance of wildlife, biodiversity, water and air quality, cultural and historical heritage – that have an elusive fiscal value (Warren *et al.*, 2008). This concept was clearly set out in the Millennium Ecological Assessment (2005) and a UK version is currently being worked on which will be published in 2010. Such a model can have both positive and negative impacts on agriculture and the environment. The assumption that an increase in an agricultural product, for example beef or dairy will increase grassland and hill pasture and thus improve biodiversity is not essentially accurate, and conversely the idea that removing an agricultural product, again beef or dairy, will improve the integrity of the environment is also misleading. With respect to beef and dairy production it is how these items are produced that has the greatest impact on the environment, and conversely removing traditional grazing from European uplands results in abandonment and subsequent change in coevolved ecosystems, which are now recognised as High Nature Value Farmland (EEA Report, 2004). Thus

Table 1.1 Evolution of agricultural and agri-environment policy in Europe

Year	Scheme	Objectives	Comments
1975	Less favoured area (LFA)	A subsidy scheme aimed at supporting farmers in highly disadvantaged areas, such as mountains and moorlands.	57% of utilised agricultural area in the EU is classed as less favoured area, however not all farmers receive LFA payments.
1987	Environmentally sensitive area (ESA) scheme	To encourage farmers to adopt agricultural practices which could safeguard and enhance semi-natural habitats in sensitive areas.	Payments were only available to farmers in environmentally sensitive areas. Now closed and superseded by the environmental stewardship schemes.
1991	Countryside stewardship (CSS) scheme	Payments made to farmers to adopt measures to conserve the semi-natural environment.	Open to farmers outside environmentally sensitive areas.
2005	Single payments scheme (SPS)	To decouple production and link payments to environmental protection through adherence to cross compliance.	Subsidy payments linked to area and environmental protection. Scheme is biased towards large estates and disadvantageous to small mixed farms. Could lead to further loss of landscape heterogeneity.
2005	Cross compliance (CC)	To regulate the application of good agricultural practice and maintain land in good agricultural and environmental condition.	Statutory management requirements help to reduce the impact of farming on the environment through mechanisms that help to reduce soil erosion.
2005	Agri-environment schemes (entry level scheme, organic entry level scheme and higher level scheme)	Three linked schemes aimed at targeting habitat and species conservation in the farmed environment.	–

Table 1.1 (*continued*)

Year	Scheme	Objectives	Comments
2005	Entry level scheme (ELS)	Aimed at increasing the area of the farmed environment under measures to protect and enhance farmland habitats. Modest payments that are linked to points for habitat protection.	Open to most farmers in Europe who agree to enhance and/or maintain farmland habitats, examples include hedgerow and watercourse management. Scheme may not actually achieve high impact on the conservation of specialist species but will help to reduce excessive degradation of common farmed habitats.
2005	Organic entry level scheme (OELS)	To increase the area of organic agriculture.	–
2005	Higher level scheme (HLS)	To target payments for advanced conservation effort.	Scheme is competitive and implementation is targeted to areas of conservation concern such as Site of Special Scientific Interest SSSI. Scheme allows for the development of special projects that can enhance specialist species. Very complicated scheme and requires farm environment plan.
2010	Uplands entry level scheme (UELS)	Aim is to support traditional farming methods in upland areas in England. The objective is to move away from compensatory payments made through hill farm allowance to a scheme that targets payments for maintaining and improving the upland landscape and environment.	A new scheme aimed at supporting farmers in marginal environments that have inherent high nature and historical value.

(*continued overleaf*)

Table 1.1   (*continued*)

Year	Scheme	Objectives	Comments
2010	Electronic identification (EID)	A mandatory and regulatory policy development aimed at electronic identification and recording of sheep movements in Europe. Objectives is to improve tractability of sheep and goats with respect to disease control.	Very controversial in the UK as high levels of animal biosecurity and market focused tractability has reduced issues associated with many infectious diseases. Cost are borne by the producer and implementation of this scheme is believed to lead to further reductions in sheep numbers in the UK uplands and therefore potentially lead to further encroachment of invasive weeds in upland areas (Chapter 3). Potentially antagonistic to UELS.

Source: www.naturalengland.org.uk, accessed 12 January 2010.

the policy makers in Europe have a demanding challenge in that they have to develop policies that support food production but simultaneously support environmental protection. Good examples of such policies are the implementation of the EU Water Framework Directive (Chapter 5) and Nitrate Vulnerable Zones (NVZs); these policies are ensuring that farmers across Europe undertake their management practices with the objective of improving water quality in all member states, and in the UK the NVZ aims to reduce nitrogen loading in fresh water to a maximum of 50 mg/l.

In the USA, the situation is significantly different, and traditionally US policy has treated agriculture and conservation as different entities, such that an increase in agricultural products is at the expense of conservation and the environment, that is a conflict. Historically in the US the approach to wildlife conservation has been to designate large areas of land as Wildlife Reserves or National Parks and remove agriculture from these areas. A good example is the development of the Shenandoah National Park in the Appalachian Mountain range, which was originally cleared by European settlers of forest and converted to agriculture, but is now largely devoid of agriculture and has been designated as a Wildlife Reserve (Crandall, 1990). However, this has numerous problems as large areas of land are scarified to promote modern intense agricultural production (Jackson and Jackson, 2002) and, of course, numerous

Table 1.2 Evolution of agricultural policy in the United States

Year	Program	Objectives	Comments
1954	Agricultural Act re-establishes flexible price support	Support US farming sector	An integral component of US Farm Bill which is the primary agricultural and food policy tool. Farm bill is modified every five years.
1956	Soil Bank Act	To retire land producing basic commodities that were surplus.	Served as a model for conservation reserve program
1985	Food Security Act	Reduce government farm support, promote export and establish conservation measures.	Origins of the wide reaching conservation reserve program and the Sodbuster, Swampbuster Programs.
1981	Conservation Easements	To reduce sub-division and development on land of high historic, scenic and conservation value.	Land owner may donate or sell a legally binding agreement that limits land use. Original activity, farming and forestry may continue. Income tax deductible.
1985	Conservation Reserve Program	Encourage farmers to convert highly erodible cropland or other environmentally sensitive areas to vegetative cover.	Cost sharing and land rental payment programme. As commodity price increases land may be withdrawn from CRP as agreements end.
1990	Wetlands Reserve Program	To encourage landowners to protect, restore and enhance wetlands on their property.	Payments based on the difference in the value of land placed under the Wetland easement. In 2002 1 075 000 acres conserved.

Source: www.nrcs.usda.gov/programs, accessed 12 January 2010 and www.nature.org, accessed 12 January 2010.

rivers, streams and semi-natural environments still co-exist in this intensely managed landscape. Consequently these surrounding and adjacent, associated habitats are subjected to environmental degradation from nutrient enrichment and downstream pollution from pesticides and animal faeces (Jackson and Jackson, 2002). The folly of this approach is now recognised and US policy is currently moving towards a multifunctional model (Warren *et al.*, 2008), indeed the conversion of and intense use of prairie grasslands for cereal production in the Pacific North West is being moderated by a 30-year programme of research and extension to mitigate against the negative impacts of soil erosion (Chapter 2).

Changes to the CAP are also aimed at meeting several international agreements such as the Convention on Biological Diversity (CBD), the first global agreement on biodiversity made at the 1992 Earth Summit in Rio de Janeiro. When the outcomes of this summit are directly applied in Europe it is immediately clear that conservation effort has to align directly with farming practice because 40% of European landscape is under agricultural production. In context with modern farming practice across Europe, application of the outcomes from Rio can be viewed as on-farm conservation of biological diversity, with farmers adopting good agricultural practice to comply with cross compliance regulations that underpin the single farm payment subsidy and fiscal support from targeted EU agri-environment policy (Table 1.1). These schemes financially support farmers in their efforts to improve species diversity and the integrity of on-farm habitats and as these schemes evolve there is the genesis of joined-up thinking. Advanced schemes such as the Higher Level Agri-environment Scheme (HLES), are designed to support farmers where effort is targeted towards vulnerable habitats and species which are often gazetted under the 1992 EU Habitats Directive, a policy that evolved as a direct consequence of the 1992 Earth Summit in Rio de Janeiro. The HLES scheme will support special projects that are aimed at increasing the abundance of rare species such as the cirl bunting and the arable flora. However there are some fundamental flaws in many of these schemes, for example the biology and locations of rare plants are not always known by project officers and, unless the farm advisors are alert to this issue, many rare plants could disappear unwittingly. The operation of the policy assumes such knowledge exists.

1.5   Further afield

Developments in agricultural and environmental policy in other developed countries differ from those outlined above and in many ways are not focused on providing farmers with financial incentives for specific conservation projects, but are more directly targeted to issues associated with the security of food production systems. In Australia the focus is on water, soil salinity and weed control, in New Zealand there is a strong focus on biosecurity with particular attention to preventing the importation of exotic pests. Nevertheless the New Zealand Ministry of Agriculture and Forestry (MAF) support sustainable agriculture through the Sustainable Agriculture Facilitation Programme, which cooperates with land users and other government and local authorities in efforts to maintain and enhance the sustainability of New Zealand agriculture. Since 1988/1989, Resource Management Grants have been provided for projects focused on flood defence and for cost-sharing efforts in projects aimed at soil conservation (www.maf.gov.nz – accessed 9 January 2010). However, this is not to say New Zealand does not commit considerable financial effort to wider conservation issues in their country. In 1987 the Department of Conservation was instigated and became responsible for 17 000 parcels of fragmented public estate that was legally protected for the purpose of conservation (Molly and Potton, 2007), and extends to 8.1 million ha of land.

In Canada the 2003 Agricultural Policy Framework has recognised the relationship between biodiversity and agriculture and has established a financial programme to support farmers in developing Environmental Farm Plans (EFPs) and environmental monitoring to identify vulnerable areas, which can be supported via the EFPs.

1.5.1   Developing countries

In many developing countries agri-environment policy is uncommon and, quite understandably, people in these countries face an enormous struggle for daily essentials that many of us living in economically developed countries take for granted. Indeed many of the problems facing people in developing countries could be alleviated by the development of a self-sufficient agricultural system that provides a secure supply of staple food products and an export surplus of exotic goods. Such a development could help to reduce the volume of bush meat taken from native forests, estimates of wildlife harvest are alarming, in the Malaysian state of Sarawak, on Borneo, 2.6 million animals are shot annually and consumed as bush meat, and in the neighbouring state of Sabah over 100 million animals are shot annually for bush meat (Bennett, 2002). In the forests of Africa and the neotropics 5 million tonnes of bush meat is consumed annually (Fa, Peres and Meeuwig, 2002). These figures are disquieting and when combined with the continued reduction in natural forest area it does not take a genius to calculate that these wild animals are facing a mass extinction event.

How can policy and agriculture work with wildlife rather than ignoring it? In the developed world some marginal improvements are being implemented but in the developing world how can the trade in bush meat be stemmed and how can the inappropriate conversion of forest to land producing global commodities such as oil palm be halted? Particularly when much of the woodland in the developed world made way for modern industrialised systems of agriculture, that is do as we say rather than as we did.

The answer is simple, fair and equitable trade, but the reality is probably unattainable in our modern fiscal society. This is not to say there is no movement towards a solution, because many international organisations have developed a number of policies to aid agriculture and forestry in developing countries. The International Tropical Timber Organisation (Chapter 4) has developed a range of policies and associated funding streams to improve the livelihoods of forest peoples around the world by promoting sustainable forestry practice. Another example of international agreement in fair trade is being developed by the Organisation for Economic Co-operation and Development (OECD). A number of policy drivers have been developed to open up trade routes for agricultural products from developing countries into the rich developed countries. The most wide reaching impact is that agricultural policy in developed countries must align with OECD rules on fair trade. These rules, known as '*Traffic Light*' rules (OECD, 2000) are based around limiting government subsidies to growers that directly distort trade and there are three categories:

Amber Box is where World Trade Organisation (WTO) members are committed to reducing domestic subsidies that distort the production of commodities that may impede the development of world trade in agricultural products from Emerging and Transitional Economies (ETEs).

- **Blue Box** includes any support that would be included in the amber box but in this scenario the support has limits on production or crops are grown over a fixed area or a fixed number of livestock, in which case there are no limits on spending in the Blue Box category and examples include compensation payments in the EU and deficiency payments in the US.

- **Green Box** includes domestic subsidies that do not distort trade and thus domestic subsidies are targeted at growers decoupled from production. Green Box programmes can support environmental protection and therefore EU agri-environment schemes and the Conservation Reserve Program (CRP) of the US is largely considered Green Box (Warren *et al.*, 2008), but some WTO members disagree with this view and argue that agri-environment schemes and CRP can lead to a distortion in trade (Warren *et al.*, 2008).

How can these rules aid people in ETE countries and simultaneously support wildlife conservation? The answer lies in opening up trade in agricultural products, for many ETE countries international trade in agricultural markets has remained static for over 20 years (OECD, 2000). Consequently, the development of employed and settled rural communities in ETE countries has been impeded, leaving many people vulnerable to starvation and hence reliant on bush meat. Agriculture is the largest employer in poor countries employing about 60% of the workforce (Bolton, 2008) and one example of international co-operation is the Economic Partnership Agreements (EPAs), which are regional trade agreements between the EU and African, Caribbean and Pacific countries (ACPs) and are aimed at instigating sustainable development (Bolton, 2008). The system is not perfect but a quote from a Kenyan vegetable grower exporting his products to the EU illustrates the key issue:

> *My family gets a good income from growing green beans for export. We have to work hard but we can get regular income from a contract with the buyer. A lot of our income goes on education for the children we are also saving to build a better house.*
>
> Mwenge, cited in Emmet (2008)

Many of us will recognise the sentiments in Mwenge's quote and therefore if wildlife conservation is going to be globally effective we come back to the central tenet of our opening argument; on-farm conservation is essential for the future security of the world's biodiversity.

But legislation alone is not enough, the best conservation on farmland occurs where the mind-set of landowners whether individuals, large corporations or the public sector is attuned to the ethos of protecting the resources of which they are the current custodians. From around the world, large and small farms, from arable, to forest to coastal – from protection of a single species, to management of a whole ecosystem, we aim to illustrate many examples of farmers who have chosen to take this way of thinking on board, and who have often had to make major lifestyle changes as a result. Their words tell of the benefits of this choice and they are the people who hope to inspire others to follow their example.

References

Bennett, E.L. (2002) Is there a link between wild meat and food security? *Conservation Biology*, **14**, 921–923.

Bolton, G. (2008) Buying and selling. *Developments*, **41**, 5–9.

Crandall, H.H. (1990) *Shenandoah. The Story Behind the Scenery*, KC Publications, Las Vegas.

EEA Report (2004) High Nature Value Farmland. Characteristics, Trends and Policy Challenges, www.eea.europa.eu/publications/report_2004_1 (accessed 7 January 2010).

Emmet, S. (2008) Miles better. *Developments*, **41**, 12–13.

Fa, J.E., Peres, C.A. and Meeuwig, J. (2002) Bushmeat exploitation in tropical forests; an international comparison. *Conservation Biology*, **16**, 232–237.

Fisher, J. (1991) *A Colour Guide to Rare Wildflowers*, Constable, London.

Hogarth, P.J. (1999) *The Biology of Mangroves*, Oxford University Press, New York.

Jackson, D.L. and Jackson, L.L. (eds) (2002) *The Farm as Natural Habitat. Reconnecting Food Systems with Ecosystems*, Island Press, Washington, DC.

Molly, L. and Potton, C. (2007) *New Zealand's Wilderness Heritage*, Craig Potton Publishing, Nelson.

OECD (2000) Agricultural Policies in Emerging and Transitional Economies. Working Party on Agricultural Policies and Markets of the Committee for Agriculture. Joint Working Party of the Committee for Agriculture and the Trade Committee Report. Available http://www.oecd.org/dataoecd/33/25/1845529.pdf (accessed 10 January 2010).

Rose, R. (2004) *Working with Nature*, Grayling, Shap, Cumbria.

Warren, J., Lawson, C. and Belcher, K. (2008). *The Agri-environment*. Pp. 38 – 41. Cambridge University Press. Cambridge.

Wet Grassland Guide (1997) Managing floodplains and coastal wet grasslands for wildlife, p. 64. Royal Society for the Protection of Birds. Sandy, Bedfordshire.

Wilson, P. and King, M. (2003) *Arable Plants: A Field Guide*, English Nature and Wild Guides Old Basing, Hampshire. Available at http://www.arableplants.fieldguide.co.uk/ (accessed November 2009).

http://www.naturalengland.org.uk (accessed 12 January 2010).

www.defra.gov.uk (accessed 12 January 2010).

www.nature.org (accessed 12 January 2010).

www.nrcs.usda.gov/programs (accessed 12 January 2010).

www.plantlife.org.uk (accessed 6 March 2010).

www.rspb.org.uk (accessed 6 March 2010).

www.sdc.admin.ch (accessed 12 January 2010).

2 Mixed farming

2.1 Introduction

Modern lowland agriculture encompasses large-scale regional, national and international food production systems that rely on favourable, benign, climatic conditions, high soil fertility and access to abundant supplies of water. In Europe the harvested area for crops such as wheat, barley, oats and rye in 2007 totalled 99 × 10^6 ha (www.fao.org/faostat – accessed 20 December 2009). These agricultural systems occupy large tracts of low-lying land that was previously/historically forested or occupied by natural grasslands or semi-arid grasslands. In the UK the wildwoods were cleared by early settlers (Rackham, 2006) over many centuries. However, more recently the conversion of lowland ecosystems, such as the peat swamp forests in Malaysia, for oil palm and rice production, have occurred within the last 50 years. This rapid conversion has led to large-scale displacement of wild populations due to wholesale destruction of their habitats.

Historically in the UK, lowland agriculture was based on a model of mixed farming, where arable crops were grown in rotation with grassland and fodder crops, which were accompanied by livestock systems such as small-scale dairy, beef and sheep production. Farmers would also grow a range of alternative food crops to meet local market trends, such as fruit, hops and fibre crops like hemp, the ecological impacts of these systems were low, the exception being the persecution of large dangerous mammals and raptors, which were hunted to extinction in the UK. Low input mixed farming, which dominated lowland UK agriculture until the beginning of the Second World War, resulted in a heterogeneous landscape. This landscape was characterised by small fields bounded by hedgerows composed of native tree and shrub species, connecting fields to woodland, which were often coppiced, resulting in further landscape heterogeneity and dynamic floristic and faunal communities. Field systems often had ditches associated with them which would have drained into the local river system and fields in the river floodplains would have been characterised by wet grasslands which experienced annual flooding, offering winter refuge for wading birds such as the lapwing (*Vanellus vanellus*, Species Box 3.3). Limited areas of the UK still exhibit this kind of landscape such as the Pulborough and Amberley Wildbrooks. This holistic lowland agricultural landscape offered various levels of niche differentiation resulting in a wide range of biodiversity, starting at the field scale

Introduction to Wildlife Conservation in Farming   Edited by Stephen Burchett and Sarah Burchett
© 2011 John Wiley & Sons, Ltd

where vegetation structure changed with the rotation of crops such as cereals, fodder crops, soft fruit, orchards and unimproved pasture which gave rise to structural diversity and ephemeral localised habitats. This was accompanied by changes in local topography and adjacent field boundaries, which resulted in regional heterogeneity (Benton, Vickery and Wilson, 2003) and the evolution of ecological associations between lowland agriculture and natural floral and faunal communities.

Cereals underpinned the lowland cash crop and in the UK lowland arable agriculture has an ancient origin, with fossil records of cereal pollen dating from 8000 years ago on the isle of Arran (Wilson and King, 2003). Cereal production is implicitly linked with the rise of human civilisations (Bahn, 2006; Wilson and King, 2003) and spread from early cultivations of wild forms of wheat such as Einkorn (*Triticum monococcum* L) and Emmer wheat (*Triticum dicoccum*) and two-row barley (*Hordeum distichon* L); and legumes such as *Pisum elatiu*. Records of cultivation date back some 12 000 years in eastern Turkey and Northern Iran (Wilson and King, 2003). The earliest domesticated cereal grains date back to c11 000 BC and originate in Syria (Bahn, 2006). The expansion of cereal grains into the UK was initially restricted to chalk and limestone plateaus of the Wessex Downs and the Cotswolds (Wilson and King, 2003) where the loamy calcareous soils were sufficiently deep and fertile to support regular arable production. Expansion of lowland farm systems was initially slow and relied on maintaining soil fertility through the use of three course rotations and inputs of farmyard manure (FYM). These lowland systems prevailed in a more or less sustainable industry for many centuries until the development of modern inorganic fertilisers and synthetic pesticides from the 1960s. These are modern developments, where the combination of expanding human populations plus the demand for increased yields, and issues associated with food security as a result of global conflicts that occurred in the early part of the twentieth century.

The ecological consequence of this long history of lowland agriculture, based on a system of, low-input low-output, is that native flora and fauna have had time to adapt to agricultural operations, such as ploughing, sowing and harvesting. In the UK, wild arable flora has evolved alongside arable cultivations and many plants that are now considered native to the UK were most certainly introduced with the seeds of food crops by the Romans, such as cornflower (*Centaurea cyanus*) and corncockle (*Agrostemma githago*).

2.2   Conventional cereal crop production

Given that cereals form the backbone of lowland agricultural systems it is useful to understand the lifecycle of a cereal crop and understand the important benchmarks in crop husbandry. This knowledge is essential in order to fully appreciate the impact cereal crops have on the environment and surrounding flora and fauna.

The following section explores the lifecycle and husbandry of winter wheat and should be considered as a model for the other three cereal crops (barley, oats and rye) however there will be species-specific variations in husbandry.

Wheat, and in particular winter wheat, accounts for half of all the cereals grown in the EU (Finch, Samuel and Lane, 2002) and in 2007 wheat accounted for 57% of all cereals grown in Europe (www.fao.org/faostat – accessed 18 April 2009). The lion's share of wheat is sown as a winter crop with drilling starting in September. Wheat is a deep-rooted crop and performs exceptionally well on heavy soils, in the UK good crop performance is achieved in the eastern and southern parts of the country and on the chalk of the Downs in Sussex.

2.3   Life cycle

Cereals are known as annual crops, in that they are sown and harvested in one year and wheat, like all cereal crops, is classified as either winter wheat or spring wheat. These terms refer to the time of sowing and vernalisation requirement, which refers to the need of the plant to experience a period of cold weather in order to trigger the switch from vegetative growth to flowering. Winter wheats require a reasonable period of cold to initiate reproductive growth and thus must be sown during the autumn whereas spring wheats have a minimal vernalisation requirement and are sown from January to March. The life cycle of cereals has been fully described and classified into identifiable growth stages that can be used by growers to manage crop inputs. The most frequently used description of cereal growth stage is the decimal growth stage (Finch, Samuel and Lane, 2002 and www.hgca.com – accessed 18 April 2009), which describes the important phases in the life cycle of any cereal crop (Table 2.1). The Home Grown Cereal Authority (HGCA) has carefully calculated the timing of inputs and husbandry to maximise crop performance for wheat, and conventional growers aim their husbandry to achieve prescribed crop benchmarks (Table 2.2), the objective of all crop inputs is to maximise yield. Knowledge of the crop life cycle is fundamental for targeting crop inputs to maximise yield response and minimise adverse environmental impacts. Application of certain chemicals at the inappropriate growth stage will result in poor disease control, undesirable economic loss and of course unnecessary subsequent environmental impacts.

2.4   Crop establishment

Safe sowings dates for each variety are published in the HGCA (2009) recommended lists (Table 2.3). Winter wheats can be sown from September to March but generally sowings start in mid September and aim to be completed by late October. Seed is sown into a prepared seedbed, which can be quite rough for winter wheat, but many growers use soil acting herbicides which require a fine seedbed. Under normal conditions the depth of sowing is between 2 and 2.5 cm but can be slightly deeper in dry seedbeds (Jellings and Fuller, 2003). Attention to seed rates is essential and these will vary depending on the desired target population and local field factors (soil type, seedbed condition, place in rotation and drainage) and timing of sowing. The aim is

Table 2.1 Cereal decimal growth stage key

	Description of growth	Growth stage	Description of growth	Growth stage	Description of growth
	Seedling growth		*Booting*		*Milk development (grain filling)*
GS10	First leaf through coleoptile	GS41	Flag leaf sheath extending	GS71	Grain watery ripe
GS11	First leaf unfolded, ligule visible	GS43	Flag leaf sheath just visible	GS73	Early milk
GS12	Two leaves unfolded	GS45	Flag leaf sheath swollen	GS75	Medium milk development
GS13	Three leaves unfolded	GS47	Flag leaf sheath opening	GS77	Late milk development
GS15	Five leaves unfolded	GS49	Awns just visible	–	–
GS19	Nine or more leaves unfolded	–	–	–	–
	Tillering		*Ear emergence*		*Dough development*
GS20	Main shoot	GS51	1st spikelet of ear just visible above flag leaf	GS83	Early dough
GS21	Main shoot and one tiller	GS55	Half of the ear emerged above flag leaf ligule	GS85	Soft dough
GS22	Main shoot and two tillers	GS59	Ear completely emerged above flag leaf ligule	GS87	Hard dough (thumb nail impression held)
GS23	Main shoot and three tillers	–	–	–	–
GS25	Main shoot and five tillers	–	–	–	–
GS29	Main shoot and nine or more tillers	–	–	–	–
	Stem extension		*Flowering*		*Ripening*
GS30	Ear at 1 cm (visible only on dissection)	GS61	Start of flowering	GS91	Grain hard (difficult to divide)
GS31	First node detectable	GS65	Half way through flowering, anthers visible	GS92	Grain hard (not dented by thumb nail)
GS32	Second node detectable	GS69	Flowering complete	GS93	Grain loosening in daytime
GS33	Third node detectable	–	–	–	–
GS37	Flag leaf just visible	–	–	–	–
GS39	Flag leaf blade fully visible	–	–	–	–

Source: From www.hgca.com – accessed 18 April 2009.

Table 2.2 Growth bench marks for winter wheat

Decimal GS	Description	Target date	Shoots (m²)	GAI	N uptake (kg/ha)	Total dry weight (tonnes/ha)	Grain yield
30	Ear at 1 cm	31 March	941	1.5	Negligible	Negligible	–
31	First detectable node	10 April	902	2.0	81	1.9	–
39	Flag leaf visible	19 May	655	6.2	189	6.9	–
59	Ear completely emerged	6 June	495	6.3	233	11.4	–
61	Start of flowering	11 June	460	6.3	248	12.1	–
71	Grain watery ripe	20 June	460	5.7	255[a]	13.7	–
87	Grain at hard dough	29 July	460	1.3	279[a]	19.6	–
Harvest	–	9 August	460	1.3	279[a]	18.4	11 at 15% moisture

[a]Nitrogen is not from root uptake but is redistributed N within the crop as leaf proteins are degraded and N is transferred to the grains.
Source: The Wheat Growth Guide, www.hgca.com – accessed 18 April 2009.

to achieve a uniform crop with a target population between 250–350 plants/m² in autumn sown crops and 150–300 plants/m² in early spring, thus winter wheat seed rates range between 100 and 250 kg/ha and spring wheats between 170 and 220 kg/ha. The concept of seed rates is important as very low final plant populations, of a cover less than 100 plants/m² give poor weed control whereas very high populations, above 350 plants/m², give rise to a higher incidence of fungal disease. A final step in crop establishment is seed dressings, this is where the seed is pre-treated with a fungicide and insecticide compound to protect young vulnerable plants from disease such as bunt, *fusarium*, *septoria* and loose smut; and insect pests such as wireworm and aphids. In cloddy seedbeds molluscs such as slugs and snails are problematical and methiocarb slug pellets are applied to the seedbed.

2.5 Nutrient requirements

The aim of this section is to illustrate to the student typical fertiliser rates and timing of fertiliser applications, required by winter wheat, for a target yield of 7–8 tonnes/ha. The rationale for this section is that the reader needs to appreciate the economic and environmental implications of crop nutrition and thus be able to critically evaluate the role of crop nutrient management in on-farm conservation.

Table 2.3 Recommended varieties of wheat

	Bread wheats		Biscuit wheats		Feed wheats	
	Soltice	Hereward	Claire	Riband	Duxford	Brompton
Region	All UK	All UK	All UK	North UK	All UK	East UK
UK treated yield (% control = 10.1 tonnes/ha)	100	91	99	No data	106	104
Grain quality						
Endosperm texture	Hard	Hard	Soft	Soft	Hard	Hard
Protein content (%)	12.3	13.2	12.0	11.8	11.4	11.7
Hagberg falling number	272	238	242	200	266	219
Specific weight (kg/hl)	78.4	79.0	76.3	74.3	76.2	73.6
1000 grain weight (g)	50.5	48.4	47.6	53.0	No data	50.0
Zeleny volume	51.0	58.1	24.1	No data	No data	No data
Fungicide treated grain yield (% treated control)						
UK (10.1 tonnes/ha)	100	91	99	No data	106	104
Dry east region (10.1 tonnes/ha)	100	92	99	No data	106	104
Wet west region (10.5 tonnes/ha)	100	90	99	92	106	102
North (10.1 tonnes/ha)	98	No data	97	99	105	103
Untreated control (% of treated control in comparable test)						
UK	80	76	78	71	82	82
Agronomic features						
Resistance to lodging (no PGR)	8	8	6	8	8	8
Resistance to lodging (with PGR)	9	9	7	7	9	8
Height without PGR	95	89	89	No data	93	84
Resistance to sprouting	7	6	5	6	No data	6

Table 2.3   (*continued*)

	Bread wheats		Biscuit wheats		Feed wheats	
	Soltice	Hereward	Claire	Riband	Duxford	Brompton
Region	All UK	All UK	All UK	North UK	All UK	East UK
Disease resistance						
Mildew	5	6	4	6	6	3
Yellow rust	9	5	9	6	7	8
Brown rust	4	5	5	3	5	7
Septoria nodorum	6	7	8	4	5	6
Septoria trittici	5	5	6	3	5	5
Eyespot	5	4	6	6	6	6
Fusarium ear blight	6	5	7	6	6	5
Latest sate sowing data	End January	End January	End February	End January	No data	End January

Source: www.hgca.com – accessed 18 April 2009.

Crop nutrition is complicated and depends on the final market, climate, soil type, previous crop and applications of FYM and in recent years the introductions of Nitrate Vulnerable Zones (NVZs) in Europe has introduced upper limits for nitrogen, in order to reduce nitrogen loading in watercourses and ground water. Typically winter wheat where straw is incorporated will remove from the soil 7.8 kg/ha of phosphorus (P_2O_5) and 5.6 kg/ha of potassium (K_2O) for each tonne of grain yield per hectare. Where straw is baled and exported from the farm 8.6 kg/ha or P_2O_5 and 11.8 kg/ha of K_2O is removed (Finch, Samuel and Lane, 2002). Nitrogen is more complicated and rates depend on the soil nitrogen supply (SNS) index (Defra, 2000). SNS is the amount of nitrogen in the soil that becomes available to the crop from establishment to the end of the growing season and is calculated from the following equation:

$$\text{SNS} = \text{Soil Mineral Nitrogen} + \text{estimates of total crop nitrogen content} + \text{mineralisable nitrogen}$$

Soil mineral nitrogen can be established by laboratory testing and thus SNS can be empirically established, however, in practice SNS is established by field assessment using published SNS tables (Defra, 2000). Indeed crop nutrition is fully documented and tables of recommended rates are published, such as RB209 (Defra, 2000) and these tables are used to design appropriate nutrient regimes for all field scale crops. Typical fertiliser rates for winter wheat are reproduced here (Table 2.4).

The timing of nutrition is fundamental and should be targeted to maintain crop growth and green area index (GAI), which is a visual assessment of canopy size expressed as a ratio of green area to the ground area occupied (Wheat Growth Guide,

Table 2.4 Fertiliser rates for winter wheat

Nitrogen rates for winter wheat, target yield (8 tonnes/ha)	SNS						
	0	1	2	3	4	5	6
Soil type	*kg/ha*						
Light sandy soils	160	130	100	70	40	0–40	0
Shallow soils over chalk	–	240	200	160	110	40–80	0–40
Deep and medium clays	–	220	180	150	100	40–80	0–40
Shallow soils over rock (not chalk)	–	–	–	120	80	40–80	0–40
Deep fertile silts	–	–	–	120	80	40–80	0–40
Organic soils	–	–	–	–	–	40–80	0–40
P or K index (all soil types)	**0**	**1**	**2–**	**2**	**2+**	**3**	**4 and over**
Winter wheat target yield (8 tonnes/ha) where straw is incorporated	kg/ha						
Phosphate (P_2O_5)	110	85	–	60 M	–	20	0
Potash (K_2O)	95	70	45 M	–	20	0	0
Winter wheat target yield (8 tonnes/ha) where straw removed							
Phosphate (P_2O_5)	120	95	–	55 M	–	20	0
Potash (K_2O)	145	120	95 M	–	70	25	0

Source: RB209 and Defra, 2000.

2008). In modern commercial winter wheat crop fertiliser applications begin in late winter/early spring with the main dressing targeted to coincide with the period of exponential growth that is stem extension GS30 to GS32. If the crop is backward, applications may be applied in early February to encourage tillering. Modern fertilisers are formed of compound mixes and thus nitrogen (N), phosphorus (P) and potassium (K) can be applied together.

2.6 Disease and pest control

The increase in yields of modern cereals has partly been attributed to higher yielding varieties, improved and increased crop nutrition but significant yield gains have been achieved by improved pest and disease control. Fungicide and insecticide programmes in wheats have been carefully designed to protect the crop from a number of fungal diseases throughout the life cycle of the crop (Table 2.5).

Table 2.5   Common foliar disease of wheat

Disease	Incidence	Risk to yield	Treatment
Foliar disease			
Septorai tritici	Present and ubiquitous every year.	Very high, currently most damaging foliar disease.	Seed dressing and foliar spray.
Powdery mildew	Common in early sown crops and the North East. High occurrence in susceptible varieties.	Low yield loss than *Septoria tritici* but can reduce grain quality.	Seed dressing and foliar spray.
Yellow rust	Common in susceptible varieties and particularly in second wheats with volunteers and with early sowings.	Very high in early sown crops.	Seed dressing has little effect. Specific foliar sprays required and applied at GS 25. Clean seedbed reduces inoculum carry over.
Brown rust	Becoming more common.	Moderate risk but can be server in warm summers.	Seed dressing has little effect. Specific foliar sprays required and applied at GS 25. Clean seedbed reduces inoculum carry over.
Septorai nodorum	Regularly found in wheat crops.	High risk to yield if flag leaf and head are infected late in the season.	Foliar spray from GS 31. Need to continue to apply if flag leaf is at risk until GS 59–61.
Root disease			
Take-all	Common in second and third wheats.	Very high risk of reduced yields in following wheat crops.	Seed treatments and rotations.
Stem based disease			
Common eyespot	Common in early sown crops and second and third wheats and first wheats following a one year break.	High risk if crop lodges due to disease infection weakening the stem.	Foliar spray at GS 25. Good rotation policy. Clean seed bed.

2.7   Weed control

Weed control in cereals has been developed as a result of the understanding gained from extensive research into the life cycle of the crop and target weeds. There are two main application strategies and these are known as pre-emergence and post-emergence. In the pre-emergence approach, generally the active ingredients are soil-acting compounds that prevent weed seeds from germinating or quickly kill off young plants. In the post-emergence approach compounds are active as either a contact herbicide or a systemic compound that is translocated round the plant. Another key issue is selectivity of action and compounds may be either broad spectrum, thus controlling both monocotyledons such as annual meadow grass (*Poa annua*) and dicotyledons species such as docks (*Rumex spp.*) and Glyphosate compounds are widely used as broad-spectrum herbicides, or selective for dicotyledons and/or monocotyledons. Herbicide formulations from the aryloxyphenoxy propionates (fops) and the cyclohexanediones (dims) families contain Acetyl coenzyme A carboxylase (ACCase) inhibitors which effectively kill grass weeds, because the ACCase of grasses are sensitive to the active ingredients in these herbicides but the ACCase of broadleaf crop plants are not. One consequence of widespread use of ACCase inhibitors is the evolution of herbicide resistant grasses such as black grass (*Alopecurus myosuroides*).

In winter wheat weed control starts in the autumn with applications of pre-emergence herbicide to control autumn germinating weeds, in particular grass weeds such as black grass and annual meadow grass. Until June 2009 isoproturon (IPU) based compounds had been widely used in the UK for 30 years to control grass weeds in cereal crops. However IPU based compounds were found to present a significant risk to aquatic life because the compound readily moves through land drains entering adjacent watercourses, and thus in September 2007 the licence for IPU compounds was withdrawn and the industry given until June 2009 to adjust. Consequently in recent years agrochemical companies have developed alternatives to IPU based compounds and these new herbicides can be used for both black grass and annual meadow grass weeds and some broadleaved weeds and are appropriate for use in the post/pre-emergence weed control strategy.

In the pre-emergence programmes compounds such as flufenacet (400 g/l) is mixed with diflufenican (100 g/l) and sold as a proprietary product called Liberator®. This combination of active ingredients controls annual meadow grass at very low dose rates (0.3 l/ha) from pre-emergence to GS 13 and up to GS21, depending on seedbed conditions. Liberator® will control black grass, ryegrass (*Lolium perenne*) and brome at 0.6 lha, again from pre-emergence to GS 13 and up to but before GS21. Furthermore the product controls a number of other 'pernicious broadleaved weeds' – also known as flowering plants such as field pansy (*Viola arvensis*), common field speedwell (*Veronica persica*), ivy-leaved speedwell (*V hederifolia*), common chickweed (*Stellaria media*), mayweeds (*Tripleurospermum spp., Matricaria spp.,*

Anthemis spp. and *Chamaemelum spp.*), groundsel (*Senecio vulgaris*), field forget-me-not (*Myosotis arvensis*), red dead nettle (*Lamium purpureum*) and cleavers (*Galium aparine*) (www.bayercropscience.co.uk – accessed 24 April 2009). As with grass weed control the control of the broadleaved weeds has an optimum window (www.bayercropscience.co.uk – accessed 24 April 2009).

Post-emergence control of annual and rough stalked meadow grass (*Poa trivialis*) weeds is achieved with a combination of active ingredients such as mesosulfuron-methyl (7.5 g/l), iodosulfuron-methyl-sodium (2.5 g/l) and diflufenican (50 g/l) sold as a proprietary product called Othello®. The efficacy of this herbicide is dependent on application rates and growth stage, for example at 1 l/ha annual meadow grass is susceptible up until GS23 but at 0.8 l/ha susceptibility of annual meadow grass to this herbicide declines and effective control can only be assured up until GS13. Othello® will also control a number of broad leafed weeds as noted above but will also control volunteer oilseed rape, which is often used in arable rotations (www.bayercropscience.co.uk – accessed 24 April 2009). The above discussion is an abridged account of weed control in cereal crops but illustrates important technical advances in crop production systems that have attributed to the significant gains in yield and crop quality, but also have resulted in the loss of floristic diversity in the UK landscape.

2.8 Harvest and crop quality criteria

Generally harvest occurs during August and September but can be as early as the last week of July under good seasonal conditions. Several quality criteria must be met before the final grain can be sold (Table 2.6) and some of these variables can be controlled at harvest. Controlling grain moisture is fundamental and growers aim to harvest at 15%, this is not always achievable in wet seasons and thus grains are dried at the point of storage. Another variable that can be controlled at harvest is admixture (weed seeds, stones, etc.) as modern combines have very efficient cleaning technology. Seed cleaning has been responsible for dramatic declines in arable flora.

Table 2.6 Quality traits for wheat

Market	Moisture content (%)	Specific weight (kg/hl)	Maximum impurities (% by weight)	Protein content % (100% DM)	Hagberg falling number (s)
Bread flour	14–15	76	2	10.5–13	250 (min)
Biscuit flour	14–15	70	2	10 (max)	200 (min)
Household flour and wheatmeal	14–15	70	2	9.5% (min)	200 (min)
Feed wheats	15	72	2	N/A	N/A

2.9   Organic agriculture

Globally the interest in organic agriculture has increased dramatically since the late 1980s with the number of organic farms in Britain increasing from 100 in 1980 to 700 in 1989 (Lampkin, 2001) and by 2007 5500 UK farms were registered as organic (www.earthtrends.wri.org – accessed 12 December 2009) with a fiscal value of 2.1 billion (www.soilassociation.org – accessed 12 December 2009) These patterns in the uptake of organic agriculture are global. In the US there are 13 000 registered organic farms. The drivers for this shift in production philosophy are numerous, ranging from direct government policies, such as organic entry level schemes (OELSs) and other policy instruments aimed at targeting improved environmental considerations (Lampkin, 2001) to individual growers who have developed an in-depth appreciation of natural ecosystems and improved business opportunities. Riverford Organic Vegetables in south Devon started from a modest 1.2-ha site but has grown to a nationwide company distributing several thousand organic vegetable boxes across the UK. Such rapid growth in organic production clearly has an impact on the environment and wildlife, which is perceived to be benign by many consumers and compared to traditional conventional farming of the 1970s through to the mid 1990s organic farming is clearly an improvement. The problem is this view is too simplistic and there are both environmental gains such as increased biodiversity (Clough *et al.*, 2007; Gibbson *et al.*, 2007a) and losses like soil erosion (Samuels, 2009, Yelland Cross Farm, South Devon Organic Growers Group, personal communication) with organic farming and once again these can be mitigated by good agricultural practice.

2.10   Organic conversion

Conversion from conventional to organic status needs to be planned and in most cases will require advice from either a government advisor or an advisor from a certifying body such as the Soil Association. It is essential that conversion is fully documented and a full farm business plan is developed. A conversion plan will include details on:

- soils map and soil management;
- crops and crop rotations;
- livestock management and health plans;
- grazing management.

Adapted from (www.businesslink.gov – accessed 12 December 2009).
In Britain conversion to organic typically takes two years but there are numerous exceptions, such as established fruit orchards and perennial soft and vine fruits,

where a three-year conversion period is required. Livestock is more problematical and generally animals intended for human consumption cannot be converted, whereas animals that produce dairy products and eggs can be converted. There are several approaches to conversion and many farmers take the plunge and convert the whole business in one process. This requires considerable attention to the financial implications as there tends to be a reduction in crop yields and consequently income as the grower is constrained to selling produce as non-organic until full certification is acquired.

2.11 Soil fertility and crop rotations

The success or failure of any organic system is centred on soil fertility. As a consequence organic farms are fundamentally based around a mixed farm system with livestock as an important element providing on-farm organic manure for soil fertility. Unlike conventional farms organic farms cannot apply inorganic nutrients such as N, P and K.

Managing soil fertility is crucial and, where FYM is available, it can be incorporated back into the soil. Incorporation of a well-composted FYM will improve soil structure and provide a range of plant nutrients for crop growth (Table 2.7). FYM is an important component of any soil fertility programme but by no means the end of the story. All organic farms use a series of crop rotations to build soil fertility, which is then subsequently exploited by a cash crop, and generally rotations are site specific, however there are a number of guiding principles in the design of rotations and these principles give consideration to a number of agronomic factors, which include weed and disease control.

Crop rotations date back several thousand years and in Britain a three-course rotation based on autumn cereals > spring cereal > fallow was used for more than

Table 2.7 Nutrients in farmyard manure

FYM source	Dry matter (%)	Total nitrogen (kg/tonne)	Total phosphate	Total potash	Sulfur	Magnesium
			(kg/tonne)			
Cattle	25	6	3.5	8	1.8	0.7
Pig	25	7	7	5	1.8	0.7
Sheep	25	6	2	3	ND	ND
Poulty						
Layer	30	16	13	9	3.8	2.2
Broiler	60	30	25	18	8.3	4.2

Source: RB209 and Defra, 2002.

1500 years (Lampkin, 2001) but by the eighteenth century fallowing was omitted by the incorporation of root crops into the system and this became known as the Norfolk four course rotation:

Roots > Barley > Seed > Wheat
(where seed was either grass or a legume)

Modifications of the four-course rotation were developed in the following century until the widespread adoption of modern arable technology in the 1960s, which enabled the continuous use of cereals. These become known as monoculture cropping. Monoculture cropping relies heavily on inputs of inorganic nutrients and widespread applications of crop protection compounds such as pesticides.

In organic systems the array of crop protection compounds is limited to elemental compounds like sulfur, therefore the design of crop rotations needs to consider the control of disease and weeds and this is best achieved by alternating a mono-cotyledon crop (cereal) with dicotyledon crops such as legumes, roots and leafy vegetables.

Starting from a point of conversion from conventional to organic the key aim is to build soil fertility, which is generally achieved by sowing a two-year ley of red clover, and ryegrass (Philipps *et al.*, 2001), which is then regularly grazed, cut and mulched. Grass and red clover leys are a key component of any soil fertility strategy and can also be under-sown in subsequent cereal crops to form a catch crop or green manure. Other green manures include both winter hardy crops such as winter rye (*Secale cereale*), vetch (*Vicia spp.*) and winter rape (*Brassica napus*), and non-hardy crops such as lupins (*Lupinus angustifolia*), clover (*Trifolium spp.*), fodder radish (*Beta vulgaris*), Phacelia (*Phacelia tanacetifolia*) and sainfoin (*Onobrychis spp.*). All these crops have the potential to protect the soil from erosion, retain, and in some cases, build up soil nitrogen, improve weed and disease control and through root action improve soil structure (Lampkin, 2001). Typically they are all fast growing and are then incorporated back into the soil.

Following the soil fertility phase crop rotations become very flexible and site specific and a typical arable rotation may consist of:

- Two to three year short-term ley: > Wheat or potatoes > Potatoes or roots > Wheat > Rye or oats.

Alternatively:

- Two to three year short-term ley: > wheat > Rye or oats > Grain legume (peas/beans) > Green manure > Wheat > Barley or oats under-sown with grass ley (Adapted from Lampkin, 2001).

The above examples are based on a system with livestock. Stockless arable rotations are achievable and this is based on incorporating as many legumes in the rotation as possible:

- Field beans/peas or vetch > Winter wheat and under-sow green manure > Spring cereal and under-sow green manure > Field beans or row crop such as potatoes or carrots > Winter wheat or cereal/grain legume mix > Winter wheat or oat under-sown with green manure.

The above is a far more sophisticated rotation and relies on nutrient building phases of the legumes and green manures. A full account of rotations suitable for a number of organic systems is given in Lampkin (2001).

The above discussions on organic agriculture illustrate the complex nature of organic systems and provide only a brief account of this subject. The important point is that like conventional agriculture organic agriculture is a system of replacing natural ecosystems with a crop production system and consequently displacing numerous species. Unlike conventional agriculture organic systems are structurally more diverse and generally support a wider range of species due to their inherent structural diversity, such as increased plant diversity in areas of semi-natural habitats (Gibbson *et al.*, 2007b) and reduce levels of toxic compounds such as insecticides entering natural ecosystems by virtue of the organic philosophy.

2.12   Summary

Historically, in the UK, lowland agriculture was based on a model of mixed farming which, when compared with the modern industrial scale arable and vegetable farms seen in the English eastern counties, operated at a slightly more ecologically sustainable pace and offered a range of habitats for local flora and fauna. However during the last 50 years, the exponential rise in global populations and the improvements in agricultural technology have resulted in dramatic declines in global biodiversity as agriculture became increasingly industrialised. In the last 20 years, member states within the European community became concerned about these losses and began to examine what was required to align agriculture with the fundamental cycles of life. In 2002 they called for a halt to the loss of biodiversity by 2010. They started to turn the clock back in efforts to restore some of the historic elements of the lowland agrarian landscape. This does not mean reverting to outdated technology, quite the opposite, the successful case studies reported here rely on integrating sound knowledge of agricultural science and the application of scientific knowledge gained from ecological studies on habitats and species of conservation concern. Indeed application of the two sciences goes hand in hand. You cannot develop a sustainable lowland farming system without knowledge of habitat ecology and species biology.

The four main case studies reported below illustrate how a number of farmers are working with government agencies and conservation bodies to reduce their impact on the local environment:

Case Study 2.1, Down Farm in south Devon. This is a small mixed farm located in a marginal coastal area.

Case Study 2.2, The STEEP programme of the American northwest that is working with arable growers across a tri-state area.

Case Study 2.3, Cholderton Estate located in Wiltshire. A large mixed organic producer with extensive areas of cereal production.

Case Study 2.4, Blueberry Hill Farm, Maryland USA. A small organic mixed farm incorporating forest fruit production.

The rationale for incorporating these case studies is that they represent diversity in scale and farming practice but all have the same objectives, which is to produce high quality food while improving the local landscape and habitats for wildlife. These case studies will also be augmented with a case study focused on the conservation of arable flora in southwest England.

Case Study 2.1: Down Farm

Introduction

In the south west of England lies the large county of Devon with coastlines on the north and to the south where the climate is mild (with mean winter temperatures at above 6 °C and summer temperatures rising to 17 °C), and relatively high rainfall gives rise to a green, lush landscape throughout the year. Devon covers some 2700 square miles (around 670 000 ha). It boasts five 'Areas of Outstanding Natural Beauty' (AONB), and two national parks – Dartmoor and Exmoor – that are famous for their rocky heights known as 'tors', which extend upwards in places to over 1800 ft (600 m). In Devon, gentle rolling hills, moorland heaths and around 300 miles of coastline are the key features of this predominantly rural environment. More than half of the land in Devon is under some kind of environmental protection, which, for landowners, often means a number of layers of legislation and thus an associated set of environmental management tasks to perform on top of their regular farming work.

Down Farm

Richard and Judy Foss are tenant farmers who run Down Farm on a rocky headland on the south coast of Devon with their son Mark. It is a mixed farm of approximately 145 ha. The South West coastal path, a 630 mile pathway which runs from Poole in Dorset on the south coast to Minehead on the edge of the Exmoor National Park in Somerset on the north coast, traverses the southern part of the farm, plus there are popular beaches nearby, so public access is high on their list of management issues. The topography of the land that they farm also presents a number of management challenges.

Topography, Geology and Geography

Topographically it's all here! Many of the fields are relatively flat as they form part of a plateau of steep sided hills giving rise to steep sided stream valleys running down to the sea. The southern fields on the farm consist of undulating land with maritime cliffs reaching 90 m in height with many other rocky outcrops. These cliffs do not extend right to the sea but culminate in shore platforms known as softhead deposits (Figure 2.1). These head deposits consist of pebbles, gravel, sand and clay, which are particularly soft and erode easily via wave action and they provide particularly important habitats for mining bees (Figure 2.2) such as the long-horned bee (*Eucera longicornis*), which is parasitised by the cuckoo-bee (*Nomada sexfasciata*) (Stubbs, 1994).

Shore platforms composed of softhead deposits

Figure 2.1 A section of the SSSI illustrating the shore platforms composed of soft head deposits. (Photo: Stephen Burchett)

The commercially suitable soil largely consists of a medium clay loam of varying depths from approximately 30 cm on the lower slopes of fields to zero where bare rock emerges, which is largely composed of Lower Devonian schists (Hesketh, 2006) and Mica-schists (Cove, 2006), which dominate the parent rock of the maritime cliffs. These schists are composed of numerous other minerals including Quartz-Mica and Quartz-Spessartine-Almandine (Cove, 2006) and gneiss which erode giving rise to localised neutral to mildly alkaline soil horizons,

Figure 2.2   Mining bees in soft head deposits. (Photo: Stephen Burchett)

that are often colonised by plants with an affinity for calcareous soils, such as the Carline thistle (*Carlina vulgaris*) which can be seen growing in the coarse grasslands adjacent to the coastal footpath.

Business Structure

Down Farm typifies the type of mixed farming system that can be found in Devon and is dominated by a grassland production system. This supports the livestock comprising 150 breeding ewes and a modest herd of beef suckler cattle. The ewes are a hybrid mixture of Northern mule × Masham × Suffolk which are hardy sheep that produce 1.5 lambs per annum on average and perform well on low input grasslands. The beef stock comprises a mixture of pure Aberdeen Angus and Aberdeen Angus × Limousin producing a hardy breed with good commercial value but also with the ability to graze the rough pasture that is found here.

The 145 ha are divided into three main cropping systems; 25 ha of semi-natural grassland dominate the southern part of the farm near the cliffs (Map 2.1). This area is part of a Site of Special Scientific Interest (SSSI), which runs from Prawle Point to Start Point. Another 48 hectares are in permanent pasture, with two fields entered into a Countryside Stewardship Scheme (CSS) and 29 ha in short-term leys used in the crop rotation system. The remaining 43 ha are in cereal production.

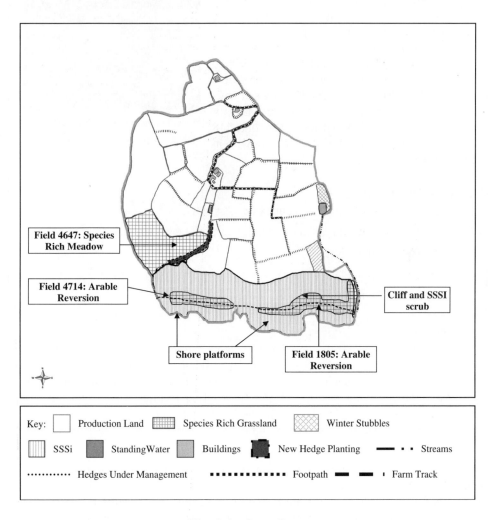

Map 2.1   Down Farm

Around 40 ha of land is currently in cereal production using the following rotation, winter wheat > winter barley > spring barley > grass ley. Six hectares of spring barley form part of a special project to enhance over-wintering stubbles for farmland birds such as the cirl bunting funded by CSS (Species Box 2.1). Successful cereal production is very challenging on the shallow and droughty soils that predominate on Down Farm; this is further exacerbated by the prevailing saline winds.

Species Box 2.1: Cirl Bunting (*Emberiza cirlus*)

Profile:

Once common in Southern Britain, the cirl bunting underwent a serious decline in the 1970–1980s due to changes in farming practices. Adults feed mainly on arable weed seeds and leftover cereal seeds in winter stubbles. This habitat has declined due to the use of winter sown crops. The chicks feed on insects such as grasshoppers, which have declined due to modern grassland management. A survey in 1989 indicated that there were only 118 breeding pairs left in Britain, of which 114 were in South Devon.

Photo: Cirl Bunting – © Edward Flatters 2006

Conservation Status:

1. Schedule 1 of the Wildlife and Countryside Act 1981.

2. On RSPBs (Royal Society for the Protection of Birds) Red List.

3. On the EC Birds Directive.

4. BAP Species (Action Plan in place).

5. In Appendix 2 Bern Directive.

Current Status:

Still confined to the South Devon area in England, there are now nearly 700 breeding pairs due largely to the efforts of the RSPB working closely with farmers. Due to their specific feeding requirements and breeding habitat (dense hedges and gorse bushes), protection and recovery of the cirl bunting requires mixed farmland with winter fallows.

The current crop rotation is 12 ha sown to winter wheat (*Triticum aestivum*) for seed production, 12 ha of winter barley (*Hordeum vulgare*) for livestock feed and 16 ha of malting barley. Typically the soils are tested on a biannual basis across the farm for pH, phosphorus, potassium and magnesium (Table 2.8) with nitrogen rates being calculated from previous crop and soil characteristics using RB209 (Defra, 2000). A characteristic fertiliser programme illustrates the attention to agronomic detail with nutrients being applied at specific growth stages in the life cycle of the crop (Table 2.9). Rates are adjusted if FYM is applied during cultivation. A balanced application throughout the life cycle of the crop ensures crop yields and quality standards with minimal detrimental impact on the environment. This is achieved due to reduced volatilisation of ammonia to the atmosphere and leaching of nitrate into neighbouring watercourses. The growing crop will generally take up a targeted application so long as there is sufficient soil moisture.

Table 2.8 Example of soil nutrient status for two sites at Down farm

Location	pH	N	P	K	Mg
Start breach	6.2	1	2	1	2
Big broadhayes	6.4	1	3	2+	2

Calculation of nitrogen index assumes a previous crop of cereals.

Table 2.9 Fertiliser recommendations for the two sample sites at Down farm

Location	N	P	K	Mg
Start breach	220	70	105	0
Big broadhayes	220	20	55	0

Rates assume that the straw removed. Where the straw is incorporated, rates can be reduced according to RB209.

Soil cultivation and seedbed preparation is based on a rotation of traditional mouldboard plough and minimum tillage technology depending on soil characteristics and season. The practice of rotating cultivating techniques is to avoid creating a soil pan due to continued use of the plough or compacting the topsoil.

Pesticide application is managed by regular crop inspection; the following fungal pathogens provide reoccurring problems at Down Farm; on wheat *Septoria tritici, S. nardorum*, brown rust (*Puccinia triticina*), mildew (*Blumeria graminis f. sp tritici*) and eyespot (*Oculimacula acuformis* and *O. yallundae*) (HGCA, 2004). On barley, the common problems are leaf blotch (*Rhynchosporium secalis*), net blotch (*Pyrenophora teres*), mildew (*B graminis f. sp hordei*), brown rust (*P hordei*) and halo spot (*Selenophoma donacis*), which is unique to the southwest coast and west Wales and infects the all important flag leaf once it emerges. All these fungal pathogens are controlled using a targeted fungicide programme prepared by a crop agronomist.

Barley yellow dwarf virus (BYDV) is transmitted by several species of aphids including (*Rhopalosiphum padi*, and *R. maidis* and *Sitobion avenae*) and is a common problem in the south west of England, the pathogen over-winters in the *poaceae* family that are abundant in the adjacent hedgerows. BYDV is controlled by applications of insecticide such as Permasect™ at 250 ml ha during the winter months after effectiveness of seed dressing has passed.

Water Courses

There are two streams on the farm which both exit to the sea, one at Lannacombe beach and the other at Great Matiscombe beach. Both beaches are used extensively during the summer by day-trippers and holidaymakers and form part of the local AONB. These streams are protected from agricultural operations and nutrient run-off by the streamside hedge planting and application of good farming practice. The farm also has two ponds that provide important source of standing water for the local bird populations.

Market

Most of the cereals are sold to a local grain merchant from whom they purchase fertilisers and sprays. The stock goes to both a local dealers and to independent dealers, with a possible move in the near future into niche markets. Niche markets often work well for smaller farmers. Unable to compete with large farms with high-yield, high-turnover of stock, the small farm is in a better position to supply local markets with something a bit extra special. This also gives rise to an opportunity to conserve some of the older and more obscure stock breeds thus increasing the gene pool and benefiting the future of the meat trade.

With a large hectarage of grassland and stock to feed all year round, making silage for their own use is essential in order to reduce winter-feed costs. Down Farm gets in one, two or three cuts a year depending on conditions and the years' rotational sequence. Yield is around 400–500 tonnes per annum.

Down Farm also runs a bed and breakfast business. With both pleasant and comfortable rooms in the farmhouse, and a large barn for use by big family groups or often groups of sub-aqua divers, this supplements the farm's income during the holiday season.

Wildlife and Conservation

Down Farm and the surrounding environment is blessed with an outstanding landscape character composed of a patchwork of rolling fields and hills, deep valleys, streams, hedgerows, small deciduous woodlands and a spectacular, predominantly south facing, coastal cliff and aqua marine skyscape. This diversity in landscape character supports an abundance of wildlife with many rare species (Table 2.10), which use the farm as part of their natural habitat. The farming philosophy adopted by Richard and Judy is one of mitigation and habitat protection where the farm works in partnership with organisations such as the RSPB, Natural England, AONB, Farming and Wildlife Advisory Group (FWAG) and Department of the Environment, Farming and Rural Affairs (DEFRA). The farm has a number of schemes to support conservation activity that includes land entered into Countryside Stewardship, Wildlife Enhancement Scheme (WES) and Entry Level stewardship (ELS). There is also a pasture management trial supported by the RSPB to establish management recommendations to increase songbird habitat and maintain commercial productivity.

Habitats for Wildlife

There is a diverse range of habitats on Down Farm including a small pocket of deciduous woodlands, scrub, hedgerows and other field boundaries, semi-natural grasslands, pasture, cereal stubbles, ponds, streams and maritime habitats (Map 2.1) all of which are exploited by numerous species of flora and fauna (Table 2.10). The woodland forms part of Lannacombe County Wildlife Site (CWS) (a local county wildlife designation), the woodland is dominated by stunted and gnarled specimens of English oak (*Quercus robur*) which are festooned with numerous epiphytes and ivy, all providing habitats for invertebrates and associated predators. In the past, gorse (*Ulex europaeus*) has dominated the woodland floor, but has been cut back in recent years as part of a CSS Agreement, to allow light into the understorey.

Table 2.10 Species list for down farm and adjacent SSSI

Scientific name	Common name	Comments
Flora		
Anagallis arvensis	Scarlett Pimpernel	–
Anthyllis vulneraria	Kidney Vetch	–
Armeria maritime	Thrift	–
Carlina vulgaris	Carline Thistle	–
Centaurium erythraea	Common Centaury	–
Centaurium pulchellum	Lesser Centaury	–
Cerastium holosteoides	Common Mouse-ear	–
Crataegus monogyna	Hawthorn	–
Digitalis purpurea	Foxglove	–
Erodium maritinum	Sea Storksbill	–
Euphorbia portlandica	Portland Spurge	–
Festuca caesia	Blue Fescue	–
Festuca rubra	Red Fescue	–
Geranium snaguineum	Bloody Cranesbill	–
Hyacinthoides non-scripta	Bluebell	–
Knautia arvensis	Field Scabious	–
Lathyrus nissolia	Grass vetchling	–
Lathyrus pratensis	Meadow vetchling	–
Lotus angustissimus	Slender Birdsfoot Trefoil	Field 4647 and 4714
Lotus corniculatus	Birdsfoot Trefoil	Field 4647
Lotus subbiflorus	Hairy Birdsfoot Trefoil	Field 4647 and 4714
Melittus melissophylum	Bastard Balm	–
Moenchia erecta	Upright Chickweed	–
Orobanche hederae	Ivy Broomrape	–
Plantago coronopus	Buckhorn Plantain	–
Prunus spinosa	Blackthorn	–
Pteridium aquilinum	Bracken	–
Raphanus maritimus	Sea Raddish	–
Rumex rupestris	South Devon Shore Dock	BAP, on SSSI (Map 2.1)
Scabiosa cloumbaria	Small Scabious	–
Scilla auntumnalis	Autumn Squill	–
Scilla verna	Aspring Squill	–
Sedum anglicum	English Stonecrop	–
Silene dioica	Red Campion	–
Silene maritime	Sea Campion	–
Stellaria graminea	Lesser Stitchwort	–
Stellaria holostea	Greater Stitchwort	–
Thymus drucei	Wild Thyme	Field 4647
Trifolium striatum	Knotted Clover	Field 4714
Trifolium suffocatum	Suffocated Clover	–
Ulex europaeus	European Gorse	–
Ulex galli	Western Gorse	–

(continued overleaf)

Table 2.10 (*continued*)

Scientific name	Common name	Comments
Verbascum thapsus	Great Mullein	Field 4714
Verbascum virgatum	Twiggy Mullein	–
Vicia cracca	Tufted Vetch	–
Vicia hirsuta	Hairy Tare	–
Viola arvensis	Field Pansy	Larval food plants for
Viola riviniana	Common Dog Violet	Frittillary butterflies
Lichens		
Buellia leptoclinoides	–	Nationally rare
Roccella fuciformis	–	Near threatened
Roccella phycopsis	–	Near threatened
Teloschistes flavicans	–	UK BAP priority sSpecies
Invertebrates		
Argynnis aglaja	Dark Green Fritillary	UK BAP priority species
Argynnis paphia	Silver Washed Fritillary	UK BAP conservation concern
Boloria euphrosyne	Pearl Bordered Fritillary	UK BAP priority species
Ectobius pallisus	Tawny Cockroach	–
Ectobius panzeri	Lesser Cockroach	–
Eucera longicornis	Long-horned Mining Bee	UK BAP priority species
Euodynerus quadrifaciatus	Mason Wasp	IUCN Red List – endangered 1994
Nomada sexfasciata	Cuckoo Bee	IUCN Red List – vulnerable 1994
Platycleis denticulata	Grey Bush Cricket	–
Plebejus argus	Silver Studded Blue Butterfly	UK BAP priority species
Polyommatus icarus	Common Blue Butterfly	–
Tettigonia viridissima	Great Green Bush Cricket	–
Timarcha tenebricosa	Bloody-nosed Beetle	–
Aves		
Carduelis cannabina	Linnet	UK BAP priority species
Emberica citrinella	Yellowhammer	UK BAP priority species
Emberiza cirlus	Cirl Bunting	UK BAP priority species
Falco peregrinus	Peregrine Falcon	Bird population – Amber

Other CSS work being carried out on the farm includes arable reversion, spring cereal stubbles, extensive grazing on lowland pasture and 6 and 3 m arable margins, established either by natural regeneration or sowing a prescribed seed mix. This mix must contain at least four different grass species prescribed by the terms of the agreement, which must be of UK origin (Table 2.11). The seed rate for the 6 m arable margins is set at 20 kg/ha with no one species exceeding 40% of the seed mix. Management of these perennial margins is laid out in the terms of the CSS agreement and consists of mowing the whole 6 m strip early and frequently in the first year following establishment, the aim here is to suppress volunteer cereals or undesirable flora. In subsequent years mowing 4 m of the strip next to the crop once a year after mid July leaving 2 m uncut adjacent to the hedgerow to act as a wildlife refuge. All cuttings should be removed. Treatments applied to the crop must not affect or encroach on the arable margin. Where grass leys are part of the cereal rotation, treatments applied to the ley must not affect the margin and where stock is grazed they must be excluded until late July. From mid July onwards the margins can be cut and a light-grazing regime applied. Other CSS prescriptions include restrictions on storing any material on these margins, such as bagged fertiliser and using the margins as farm tracks, again the aim is to ensure high quality undisturbed habitat for local fauna.

Table 2.11   Recommended seed mix for establishing new field margins at Down farm

Hedgerow planting		Field margins (6 m)	
Scientific name	Common name	Scientific name	Common name
Acer campestre	Field Maple	Agrostis capillaris	Common Bent
Corylus avellana	Hazel	Alopecurus pratensis	Meadow Foxtail
Crategus monogyma	Hawthorn	Anthoxanthum odoratum	Sweet Vernal Grass
Ilex aquifolium	Holly	Cynosurus cristatus	Crested Dogs Tail
Malus sylvestris	Crab Apple	Festuca ovina	Sheeps Fescue
Prunus padus	Bird Cherry	Festuca pratensis	Meadow Fescue
Prunus spinosa	Blackthorn	Festuca rubra (subspecies commutata)	Red Fescue
Rosa canina	Dog Rose	Poa pratensis	Smooth Meadow Grass
–	–	Trisetum flavescens	Yellow Oat Grass
–	–	Phleum pratense (subspecies bertolonii)	Small Leaved Timothy

The streams on the farm are protected from farm operations by new fencing and hedge planting (Table 2.11, Map 2.1), again supported by CSS agreement. The benefits of this stream side planting is significant as both the streams on Down Farm exit at popular beaches so that maintaining water quality is of paramount importance.

The extensive grazing on the lowland pasture is designed to increase the diversity of local grassland flora, in particular field 4647 (Map 2.1) in early June. Dense stands of harebells (*Campanula rotundifolia*) occur and thus grazing is managed using cattle and sheep for 10 weeks a year between April and November, aiming for a target sward height of 50–75 mm (2–3 in.) by the end of the grazing period. Of course, this is subject to weather conditions, as poaching must be avoided at all cost. Thistles (*Cirsium vulgare* and *C. arvense*), two very invasive weedy species, are controlled by cutting twice a year; first cut prior to flowering and at a minimum height of 150 mm, followed by a second cut in late summer at a minimum height of 40 mm to avoid late seeding.

Winter stubbles are another significant wildlife habitat on the farm and Richard Foss sows 6 ha of spring barley each year under CSS agreement where 1.58 ha is divided between two sites as a permanent spring crop (Map 2.1) and the remaining proportion rotated around the farm. The importance of spring barley as a conservation crop is its key role in the life cycle of cirl buntings, which forage on the spilt grain during the winter months. Cirl buntings

Species Box 2.2: Skylark (*Alauda arvensis*)

Profile:

A small songbird of 16–18 cm, mainly brown with white, and a small crest on the cap. The skylark is particularly noted for its spectacular song-flight. The males fly rapidly and almost vertically to often over 100 m whilst singing their distinct song. Also distinctive is its crouched walking position. It can be seen across the UK and in parts of Europe, Asia and North Africa. The skylark is a ground nesting bird, preferring open countryside, especially farmland. It has a fairly long breeding season and has been known to raise four broods in a year.

Photo: Skylark Chicks on Nest in Cropfield 2004. © Stephen Burchett

Conservation Status:

- Protected under the Wildlife and Countryside Act 1981.
- RSPB Red List of endangered species.

Current Status:

Skylarks have suffered rapid decline in numbers due to changes in farming practice. Nesting occurs where optimum crop conditions offer cover of 20–50 cm. In the past spring-sown cereals were planted and skylarks would raise chicks in late spring and early summer. Autumn-sown cereals are now commonly planted, and the skylarks may raise only one brood before harvesting. As a result, estimates for their decline are in excess of 75% in the last 25 years. As a result, the RSPB are managing farmland plots to encourage skylark nesting and farmers can use this same method on their land as one of the options for the Entry Level Scheme.

References

www.rspb.org.uk

Peterson, R., Mountfort, G. and Hollom, P.A.D. (1983) *A Field Guide to the Birds of Britain and Europe*, 4th edn. Pub. William Collins & Co Ltd, Glasgow.

(Species Box 2.1) are not the only beneficiaries. Many other farmland birds including corn buntings (*Miliaria calandra*), yellowhammer, skylarks (*Alauda arvensis*) (Species Box 2.2) and finches (*Fringilla* and *Carduelis spp.*) will forage on winter stubbles. However, growing a cereal crop for conservation purposes is not just a matter of sowing the seed and walking away but is a matter of first class husbandry with attention to cultivation techniques and crop husbandry.

Cultivation begins at the end of March in each year and will be either mouldboard plough or minimum tillage, seed is sown at 125–180 kg/ha. The crop is grown with the aim of achieving malting barley standards and finally the crop is harvested during July/August. Often cereal conservation crops are left standing and this has some significant detrimental issues.

The other major conservation work being conducted at Down Farm is the restoration of the maritime grasslands that form part of the Prawle Point to Start Point SSSI (UK National Grid Reference SX 741371 to SX 819381), which extends to a total area of 341.2 ha of which Richard is custodian of approximately 25 ha. This site has been classed by Natural England as unfavourable but recovering, the unfavourable status is due to the aggressive scrub vegetation that includes stands of gorse and bracken (*Pteridium aquilinum*), which dominate the entire site. Mr Foss has entered the land into a WES and has made significant progress in controlling both the bracken and the gorse. This has been achieved by a combination of cutting back the vegetation and opening up areas for grazing with cattle and sheep. The favoured breed is Aberdeen Angus, which can perform well on the poor grasses and steep banks that dominate this site.

The importance of this work cannot be overstated as this site is noted for numerous rare species including South Devon shore dock (*Rumex rupestris*) which is a Biodiversity Action Plan (BAP) Species and autumn squill (*Scilla autumnalis*) a diminutive plant, that is easily out competed by the taller aggressive scrub vegetation. Another detrimental impact of the scrub vegetation is the decline in invertebrate habitats, in particular patches of bare ground which are favoured by mining bees (*Andrena spp.*) and digger wasps (*Aphilanthops spp.*). Again, grazing by cattle will open up the vegetation and the large hoof prints will afford opportunities for burrowing invertebrates. The aim is not to eradicate all the scrub but to create a mosaic of scrub and maritime grassland and Natural England has a target of 10% scrub cover for the site, which is a particularly ambitious target and perhaps should be modified to between 20 and 30% scrub cover, given the topographic constrains of the site (Figure 2.3).

Under a CSS agreement two fields that total 8.53 ha, on the lower platforms of the SSSI have been entered into an arable reversion agreement (Map 2.1). The project has been

Figure 2.3 A section of the SSSI at Down Farm. (Photo: Stephen Burchett)

running for a number of years. The aim of arable reversion is to try and increase diversity in regional flora by converting marginal arable fields back to grassland. In order for fields to be considered for reversion there must be a number of important species recorded within the fields under consideration.

Management of these fields is laid down under the CSS agreement and options are limited by EU regulations. In practice, management is via annual grazing with either sheep or cattle and for a period of at least 10 weeks between April and November. Stocking rates must not exceed 1.4 LSU/ha. Grazing must ensure that grass growth is controlled and an average sward height between 50 and 75 mm is achieved by the end of the stocking period. Prescriptions also call for a period of continuous grazing at low stocking density rather than shorter periods at a higher stocking density. The aim here is to reduce the risk of poaching, which can occur under high stocking rates. Care must also be exercised during wet periods and wet summers which give rise to wet ground conditions (field capacity) circumstances that readily lead to poaching. Other conditions apply such as no supplementary feeding and no creep feeding of young livestock, however, mineral licks are allowed following approval of location by Defra.

However, there is some cause for concern about this project, firstly these fields have historically been used as arable sites which includes the cultivation of cereals, cauliflowers and potatoes and consequently the flora which has evolved over time is that primarily of arable sites. The life history of such flora has evolved a strategy known as a ruderal (Grime, 1974). Ruderal plants are adapted to highly disturbed habitats that are abundant in resources, such as soil nutrients, and water and a benign prevailing climate. In nature such environments can be found in naturally disturbed habitats such as the prairies and savannahs where lightning strike initiates a rapid grassland burn and consequently removes standing vegetation. This affords opportunities for germination of annuals, biennials and short-lived perennials that rapidly respond to the ample soil nutrients released by the burn. Ruderal strategy also corresponds to *r-type* organisms. Such organisms have short life spans, rapid generation times and produce copious quantities of off-spring (seed), which often has a short window of viability. Indeed many agricultural pests are *r-type* organisms.

In the context of an arable site it is clear that *r-type* organisms have dominated the soil seed bank, where the life history strategy is adapted for high disturbance and rapid exploitation of the ample resources (in this case nutrient rich soils). Many of these wild plants are exceptionally rare and are collectively known as arable flora. This flora does not exist in isolation and it is not unexpected to find a varied array of invertebrate fauna exploiting the nectar provided by species such as corn marigold (Species Box 2.3). Arguably, therefore, applying an arable reversion philosophy to these two fields is fundamentally flawed.

A walk through these two fields will quickly illustrate the point. Over the last 8–10 years, following reseeding with a suitable grassland mix, the fields have developed into excellent examples of mesotrophic grassland, with little evidence of increase in valuable and threatened flora. If these sites were originally of high value due to other criteria, for example drained wetland site or chalk grassland then arable reversion would be more appropriate. But in this example this is not the case and the historical land use of this site has given rise to the development of some very important flora that has now been damaged by inappropriate application of conservation practice. This example clearly illustrates the importance of understanding the historic relationship between farming, habitat ecology and species biology.

Species Box 2.3: Corn Marigold (*Chrysanthemum segetum*)

Profile:

Growing mainly in arable land as it favours nitrogen rich soils, this bright little plant of the daisy family (*Asteraceae*) is generally found growing in association with other arable plants in species-rich communities. Though its range is widespread, it prefers sandy soils and sand-loam soils and is therefore most common in the southwest of England. It is tough and multi-branched with single bright yellow flowers. One of the corn marigolds' most notable features is that it produces two types of seed. From the inner disc-florets they produce cylindrical un-winged seeds and from the outer ray-florets they produce broad winged seeds. The two types of seed are set at different times so this is probably a strategy to exploit different conditions and possibly different harvest times that may otherwise cause a loss of the flowers before seed is set. They do best in spring grown cereal crops.

Photo: Corn Marigold © Stephen Burchett 2009

Conservation Status:

N/A

Current Status:

The corn marigold is widespread across Britain and Ireland as well as parts of Europe. Though it has suffered some decline as a result of changes in agricultural practice, such as use of herbicides and conversion of arable land to pastureland, it is not generally considered to be under threat.

References

Wilson, P. and King, M. (2003) *Arable Plants – A Field Guide*, English Nature and *WildGuides*, Ltd Hampshire, England.

Conclusions

The benefits of managing the land for wildlife can clearly be seen when walking around Down Farm in the springtime. Amidst the sound of skylarks singing and bumblebees buzzing, a stroll down the steep cliffside reveals an incredible variety of wild plant species, including some rarities. Deep pink fumitory (*Fumaria spp.*) and a particularly rare dock species, the South Devon shore dock nestle among the stitchwort (*Stellaria holostea*), bluebells (*Hyacinthoides non-scriptus*), chickweed (*Stellaria media*) and purple and lilac violets (*Viola riviniana*), whilst bugle (*Ajuga reptans*), celandine (*Ranunculus ficaria*), speedwell (*Veronica persica*) and wild mint (*Thymus spp.*) form large patches. The subtle aroma of the wild mint, however, is overwhelmed by the coconut scent of the European gorse (*Ulex europaeus*) that stains the rocky sea cliffs in vivid yellow. At the bottom of the hill, a flat strip of pasture that was once in cultivation overlooks the rock pools on the beach and in the cliff face of these soft head deposits numerous colonies of mining bees can be found. The very fact that all this biodiversity is readily seen is a testimony to the effort Richard and Judy put into their daily working life, which can be enjoyed effortlessly by taking a walk along the southwest coastal path.

2.13 Arable Flora

The plants found among arable fields – the arable flora are annuals and or short lived perennials that have coevolved with arable cultivation and are commonly called 'weeds' by the farming community. An arsenal of herbicides has been used to target arable flora for eradication from the cropped environment, with many species now significantly threatened in the UK and further afield. The threatened nature of this group of flora has led to some important conservation efforts funded through the CSS, and the first arable area to be awarded a grant under CSS was at West Pentire in Cornwall, which is being managed by the National Trust (NT).

The NT is a major landowning organisation in the UK that has several properties in the coastal regions around the UK and has a number of very important sites in Devon and Cornwall. The first priority of the Trust is of course the maintenance and preservation of its historic houses but the NT also has commitments to UK biodiversity and is making a grand effort to preserve marginal coastal farms, one such farm is East Soar Farm in South Devon, while another site supports an arable flora project at West Pentire in Cornwall. Both of these sites are very challenging and present significant management issues but at East Soar farm an effort is being made towards the conservation of arable flora in conjunction with spring cereal crops, winter stubbles and the cirl bunting.

East Soar Farm is situated on Bolt Head near Salcombe in close proximity to Down Farm, which is situated at Start Point near Kingsbridge. Both farms are exposed to the coastal environment and are predominantly situated on soils derived from Devonian Schist (Hesketh, 2006). East Soar Farm is managed as a mixed farm under an organic system and two fields are sown to spring barley under an agri-environment agreement and managed to enhance the populations of cirl buntings and arable flora. Unlike Down Farm the diversity of arable flora is quite high at East Soar, with 56 species recorded across the two arable fields compared to the 16 species observed in the two permanent arable sites at Down Farm (Map 2.1). The composition of observed species is also important and at East Soar Farm there are good populations of corn marigold (Species Box 2.3), night flowered catchfly (*Silene gallica*) (Species Box 2.4), dwarf and sun spurge (*Euphorbia exigua; E helioscopia*) round-leaved fluellen (*Kickxia spuria*) and stinking chamomile (*Anthemis cotula*), which are all noted species due to their restricted distribution. The species compositionobserved at Down Farm is primarily common arable plants and in some cases problematical species such as bramble (*Rubus fruticosus*) and a high population of

Species Box 2.4: Night-Flowering Catchfly (*Silene noctiflora*) and Small-Flowered Catchfly (*Silene gallica*)

Profile:

Catchflys are related to the more common and familiar campions (*Silene spp.*) and are found mainly growing in association with arable crops especially in field margins. Both the species highlighted here have a widespread, low-density range in Britain; the night-flowering mostly favouring the chalk and limestone soils of the eastern and southern parts of Britain; the small-flowered mainly favouring the sandier soils of southwest England and southern Wales. They are also found in Europe. Named catchfly because of the sticky

Photo: Night-Flowering Catchfly © Stephen Burchett 2007

upper parts of their hairy stems, they are distinguished by their five notched narrow petals (especially deep notches in the night-flowering) that are white or pale coloured, and their characteristic oval seed capsule. Both are insect pollinated, though the night-flowering is pollinated only by nocturnal insects, mainly moths, as the flowers remain tightly rolled during the day. The night-flowering catchfly flowers between July and September, the small-flowered catchfly flowers between June and October.

Conservation Status:

- Night-Flowering:
  - — BAP – Species of Conservation Concern;
  - — IUCN Red List – Vulnerable;
- Small-Flowered:
  - — BAP – Nationally Scarce: Priority List;
  - — IUCN Red List – Endangered;
- Both Protected under the Wildlife and Countryside Act 1981.

Current Status:

Once widespread throughout Britain and Europe, they have suffered massive decline, especially in Britain and the northern parts of Europe due mainly to changes in farming practice. They are highly vulnerable to herbicide sprays and fertilisers, have suffered from removal of arable crops in favour of pasture, and from the loss of hedgerows.

References

Wilson, P. and King, M. (2003) *Arable Plants – A Field Guide, English Nature and WildGuides*, Ltd Hampshire, England.

annual meadow grass (*Poa annua*). Two variations in management practice account for this. Firstly there are no herbicides applied at East Soar Farm and consequently there is an increase in the diversity of flora and thus an increased probability of subsequent recruitment from the soil seedbank. Secondly, the arable site at East Soar has been managed on a rotational basis that favours the life history strategy of *r-type* species. At Down Farm the continued or permanent arable leads to an increased abundance of grass species and although these permanent arable sites at Down Farm are highly disturbed it is the application of herbicide that has favoured a higher abundance of grass species compared to the vulnerable dicotyledonous arable flora. Sadly a similar turn of events is occurring at West Pentire.

West Pentire is a small coastal arable farm situated on the exposed cliffs of the North Cornish coast near Newquay, the farm comprises 11 arable fields that extend to 16 ha. Unusually for Cornwall the site contains a soil, which benefits from windblown sand thus raising the calcium levels of the soils that are generally acid as they form over the granite backbone of the county. Adjacent to the site are fields covered with cowslips (*Primula veris*), which are annually used in the May Day celebrations held in Padstow. This plant thrives only in poor soils containing calcium.

The farm is known for its important arable flora such as Venus's-looking glass (*Legousia hybrida*), rough poppy (*Papaver hybridum*), shepherd's needle (*Scandix pectin-veneris*) and small flowered catchfly (*Silene gallica*) (Species Box 2.3). However these important plants are being out competed by competitive plant species due to the continued use of shallow cultivation, in this case disc cultivation, which fails to successfully turn over the soil seed bank, resulting in compacted surface horizons (Figure 2.4). These conditions favour aggressive and competitive weedy species such as black medic (*Medicago lupulina*), perennial sowthistle (*Sonchus arvensis*) and creeping thistle (*Cirsium arvense*). Indeed hawthorn (*Crataegus monogyna*) sapling scan be

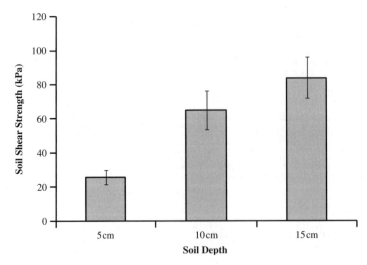

Figure 2.4 Soil compaction observed in one field system at West Pentire. Soil shear strength is a measure of the shear stress, with high values indicative of compacted soils

observed up to 3 m in from the hedgerow. There has been some limited use of glyphosate herbicide but this occurs every five years (Wilson and King, 2003) on a rotational basis and thus fails to suppress the competitive weeds in the succeeding four years when no herbicide is applied. When this practice is coupled with shallow disc cultivation, conditions that do not fully represent disturbance, then competitive species will be favoured and consequently *r-type* species will be selected against. The best way forward for this site is a return to using a mouldboard plough on a rotational basis with disc cultivation to ensure the soil seed bank is turned over and the seeds of the vulnerable arable flora are exposed to germinating conditions.

2.14   Elveden estate

In Chapter 1 we discussed different approaches to integrating farming and conservation. Many farms are too small to economically justify removing large sections of land from production in order to meet conservation demands, where this is the case the landowner needs to find ways to blend conservation and farming practices, for example maintaining cereal stubbles after harvest for the cirl bunting, highlighted in Case Study 2.1 above; or conservation grazing practices highlighted in Chapter 3. On larger farms however, it can be relevant to dedicate whole parcels of land to conservation. Some species simply cannot tolerate the disturbance caused by farming and/or their habitat requirements are too specific for survival in any but the most minimally managed land, for example the starfruit plant (*Damasonium alisma*), highlighted in Species Box 1.1 that only survives in vernal pools. Its germination was triggered by cattle trampling into the ponds to drink . . . so the reduction in herds/availability of ponds being used for drinking water . . . led to the demise of the plant and Plantlife volunteers trample in the ponds to reproduce the actions of the cattle to ensure successful germination.

The Elveden Estate is a large estate in Suffolk, England. Here both systems are in place, that is large portions of land are dedicated specifically to wildlife conservation, but also farming practice is managed sympathetically for wildlife. Elveden Estate comprises 22 500 acres (9000 ha), of which 10 000 acres (4000 ha) support protected heathland, coniferous forest and woodland shelterbelt. The remaining land is farmed for vegetables, cereals, block hedgerow planting and Christmas trees plus there is some grazing by external livestock farmers. The size of the estate ensures that there are sufficient personnel to manage both the farmland and the areas not given over to production.

The variety of bird species is particularly notable at Elveden and there are regular sightings of little owls (*Athene noctua*), hobbies (*Falco subbuteo*), redstarts (*Phoenicurus phoenicurus*) and woodlarks (*Lullula arborea*), plus the occasional sighting of the hoopoe (*Upupa epops*) in the woodlands and marginal lands and skylarks (*Alauda arvensis*) are regularly seen amidst the croplands. Most notable, however, is the stone curlew (*Burhinus oedicnemus*), Elveden boasts 25% of the population of stone curlews in Britain. These ground-nesting birds are known to favour a heathland habitat, although on the estate many of them favour the onion fields where the shallow drills and partial cover suits their needs for both their concealment and all-round lookout for predators. Of course, the onion fields are a little 'inconvenient' for the

---

Species Box 2.5: Stone Curlew (*Burhinus oedicnemus*)

Profile:

The stone curlew, sometimes called the 'thick-knee' is a medium sized wading bird with distinctive large yellow eyes. Over-wintering mainly in Eastern Africa and the Canary Islands, it is a summer migrant to breeding grounds in many parts of Europe including the UK and to South-East Asia. Breeding pairs are monogamous and the birds generally return to within 15 km of where they hatched. Breeding grounds tend to be dry stony ground with short vegetation, and sympathetically managed arable land. Stone curlews are not related to curlews and are only named thus because of the similarity of their call. They feed on worms and small invertebrates.

Photo: Stone Curlew © Jan Champion 2009

Conservation status:

1. BAP Species.

2. Schedule 1 Wildlife and Countryside Act 1981.

3. Annex 1 EU Birds Directive.

Current Status:

In recent years numbers have been increasing from an all time low of about 150 breeding pairs to in excess of 350 breeding pairs. Two strongholds: The Brecks of East Anglia, due in part to the Elveden Estate in Suffolk and to the Norfolk Wildlife Trust; and the chalk downlands in Wessex, due in part to the RSPB, English Nature, the MOD and DEFRA.

References

Elveden Estate (Personal Communications)

Elphick, J. and Woodwood, J. (2009) RSPB Pocket Birds of Britain and Europe. 2nd edn, Pub. DK Ltd, London.

---

estate farmers, Elveden is not organic so crops are periodically sprayed. Where stone curlews are nesting, the estate workers have developed a system using 'false eggs' to temporarily replace the eggs on the nests whilst spraying, the real eggs are later returned. The stone curlew parents are not put off by this practice and return to the nest to continue incubation.

The gamekeepers, in-house staff and volunteers monitor these and other birds, and plan next year to include bats and reptiles in their conservation-monitoring schedule (Species Box 2.5).

Case Study 2.2: STEEP Programme in the Pacific North West

Introduction

One of agriculture's best-known environmental disasters was the occurrence of the dust bowls of the American plains in the 1930s; in particular the southern plains that stretch across the states of Colorado, Kansas, New Mexico and Texas, but also other intensive cereal production areas such as the northern plains were affected. The agricultural practice, which gave rise to these dust bowls was the extensive conversion of prairie grasslands, by deep ploughing, for the cultivation of wheat. During the early 1930s the southern plains experienced periods of prolonged drought and this, coupled with the subsequent decline in soil organic matter led to the development of very loose and friable soils that were readily whipped up by the high winds which commonly occur in the southern plains. These devastating environmental events led to the migration of many thousands of people from the plains in search of work, food and new homes leading to civil unrest in areas such as California, as a result of migrant workers flooding the employment market. In 1934, the drought reached its zenith affecting 75% of the country and the great dust storms, called 'black blizzards' spread from the dust bowl area. During this exceptionally dry year approximately 35 million acres (just over 14 million ha) of formerly cultivated land was destroyed, a further 100 million acres (40 million ha) of contemporary cropped land lost half of its topsoil and another 125 million acres (50 million ha) were rapidly losing topsoil through wind blown erosion (Yearbook of Agriculture, 1934).

A combination of events from large-scale degradation of soil and cropping area, migration of people from rural areas to urban centres and the depressed state of global agriculture, during the 1930s, led to the development of federal policy to reduce the impact of poor agricultural practice in crop areas with fragile soils. On April 27 1935, Congress passed an act that established the Soil Conservation Service (SCS) in the Department of Agriculture. The SCS developed a wide ranging soil conservation programme that included strip cropping, terracing, crop rotations and cover crops. Farmers were paid to practice soil conservation farming techniques. The story does not end there and today in the north west there is a tri-state programme, the STEEP programme (Solutions to Environmental and Economical Problems) that aims to reduce the impact of industrial scale agriculture on soil erosion.

(STEEP)

The STEEP programme is a soil conservation initiative that extends across three major wheat production areas in the Pacific Northwest, incorporating the states of Idaho, Oregon and Washington. The aim of the STEEP programme is to reduce soil erosion losses and to reduce the subsequent pollution of watercourses to tolerable levels by implementing economically sustainable solutions.

This important breadbasket covers 10 million acres (4 046 000 ha) of cropland and four major land areas, the Columbia Plateau, Palouse-Nez Perce Prairies, Columbia Basin and the Snake River Plains (Map 2.2). STEEP was implemented in 1975, the rates of soil erosion losses were estimated at 110 million tonnes annually (>120 million tonnes annually) (Papendick et al., 1983) and, of this amount, an estimated 30 million tonnes annually were being deposited in the rivers, streams, lakes and harbours of the Pacific Northwest (Powell and Michalson, 1983) which resulted in the degradation of the natural ecosystem functions of these watercourses.

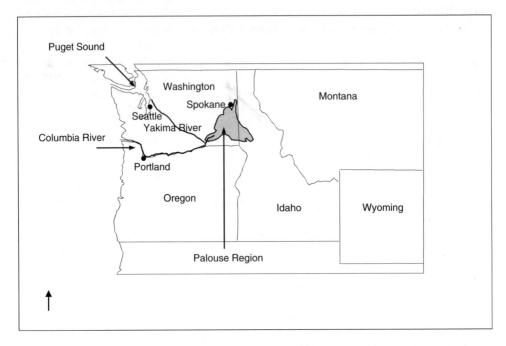

Map 2.2   Tri-state region of the STEEP program illustrating the Palouse area and the Snake and Columbia River Systems and downstream estuaries, (Map not to scale for illustration only)

With respect to water quality and degradation the project's initial focus was on the high precipitation area of the Palouse region where soils are characterised by young thin loess, deposited about 18 000 years ago (McCooll and Busacca, 1999). Originally these soils were covered in grasslands and the natural processes involved in the life cycle of grasses led to the development of a dark brown silt-loam, with a granular structure high in humified organic matter (McCooll and Busacca, 1999). This is known as the A-horizon and is the principle zone of plant rooting, nutrient cycling and biological activity, consequently the high fertility of these soils attracted the attention of arable farmers, which led to large scale conversion of these grasslands to deep cultivated wheats and other arable crops. Over time this change in land use led to significant degradation in soil structure and consequently high rates of soil erosion during the first 70 years of the twentieth century. Rates of soil losses were dramatic with 12 bushels of soil lost for every bushel of wheat produced. In metric terms, for every 27 kg of wheat produced 326 kg of topsoil was lost to erosion. Quite obviously this was unsustainable and in 1975 STEEP was established with the intention of providing answers through research, cooperation and extension.

The solution to this excessive soil erosion lay in reducing tillage operations and maintaining green cover in the winter fallows – two agricultural practices that can lead to significant crop management issues in the following wheat rotation. The regions' growers viewed conservation tillage with scepticism and consequently initial uptake of minimum tillage strategy was poor. This led the steering group of the STEEP programme down a clear path with the development of six core objectives:

1. Development of tillage and plant management.
2. Plant breeding.
3. Erosion and run-off prediction.
4. Pest control.
5. Socio-economics of erosion control.
6. Integrated technology transfer research.

Resolving issues associated with each one of the objectives was essential in establishing wide scale grower support and up-take of minimum tillage systems. Taking the first point, when a grower leaves soil fallow without residue cover (crop debris and germinating weeds) by deep ploughing s/he is effectively reducing soil borne disease inoculums, pest and weed thresholds for the next crop. However, simultaneously the grower is contributing to soil erosion by leaving the soil exposed to the elements.

Palouse Region

The Palouse region (Map 2.2) is recognised for its productivity and is a major producer of wheat, peas and lentils and the Palouse supplies more than 10% of the white winter wheat grown in the USA (USDA, 1979). Arable agriculture began in the 1880s and the region has suffered from significant losses of fertile topsoil through soil erosion since. Losses of topsoil are staggering and from 1939 to the late 1970s the region lost a total of 390 tonnes per acre (963 tonnes per ha) and 30% of this eroded soil was being deposited in lakes, watercourses and eventually reaching the sea (USDA, 1979). In many rivers small islands have developed solely as a product of soil erosion and consequently changing the natural dynamics of watercourses.

Causal Factors Leading to Soil Erosion and Types of Soil Erosion

Soil erosion is caused by a combination of factors including the topography, soil structure, prevailing climate and field operations. The Palouse region is divided into three distinct areas with recognised patterns in land use, precipitation and water erosion potential (Table 2.12). It is when activities such as ploughing, harrowing and harvesting are carried out in the fields on these young fragile loess soils, which thinly cover the slopes and ridge tops of the Palouse that soil erosion rates are most severe (USDA, 1978).

Causal factors leading to anthropogenetically induced soil erosion include the use of heavy farm machinery and constant working of the soil in combination with inorganic nutrient fertiliser in place of FYM. This leads to a significant loss in soil carbon, eventually the collapse of soil structure and finally compaction of the upper soil horizons. Such

Table 2.12 Classification of environmental and cropping zones in the Pacific North West

Zone	Description	Estimated cultivated area (000, ha)	Annual precipitation (mm)	Soil depth (mm)	Typical historic rotations	Soil organic matter (%)	Average water holding capacity (mm/m of soil)	Water erosion potential
1	Cold, moist	N/A	>400	All	Annual crop, grass (5–6 yr) /SB or DL/WW	4+	200	High
2	Cool, moist	719 830	>400	All	WW/DL, WW annually, WW/SB/DL	3–4	181	High
3	Cool, deep soil, moderately dry	276 920	350–400	>1000	WW/SP (green or dry), WW/F, WW/SC/F	2–3	165	High
4	Cool, shallow soil, dry	455 870	250–400	<1000	WW/, annual SB or after F, WW/SC/F	<1.5	148	High
5	Cool, deep soil, dry	190 485	250–350	>1000	WW/F, WW,SC/F, WW/F/F	<1.5	148	High

DL = Dry Legumes, F = Fallow, SB = Spring barley, SC = Spring Cereal, WW = Winter wheat.
Adapted from Douglas *et al.* (1999).

compacted surface horizons impede percolation of surface water through the soil profile and consequently the winter rains and the spring thaw give rise to water-borne soil erosion as surface waters run off the ridge tops and down the numerous slopes into the valley bottoms and adjacent watercourses. During the dry summer months wind-borne erosion occurs where soils are left fallow (no vegetation cover) and these erosion events become very significant moving soils into neighbouring districts and depositing surface soil onto young crops.

Technically there are several types of soil erosion and heavy winter rains on compacted frozen soils leads to sheet erosion events, where soil particles are removed from an entire surface area resulting in the loss of fertile topsoil. Heavy rains also cause rill erosion where closely spaced rills occur on the soil surface. Rill and sheet erosion are commonly observed in the Palouse region and often intense periods of winter rain on steep slopes will cause soil slips. Finally gully erosion is where run-off water from compacted land accumulates in natural channels and results in soils being washed away from the edge of the channel or stream. Gully erosion is very visible and difficult to control and results in soils being deposited on adjacent fields and once again into watercourses (USDA, 1978).

The loss of topsoil has profound financial implications for farmers and impacts on the biological and ecological integrity of the regions watersheds. In order to ameliorate these highly damaging events the STEEP programme has developed a series of on-farm conservation strategies that have reduced soil erosion from as much as 20 tonnes per acre annually to around 5–7 tonnes per acre annually. This is pragmatically accepted as a tolerable threshold without impairing financial viability of the region's farms.

Estimates of soil erosion and sediment yield have been developed for the two major classes of land use in Palouse River Basin. Sediment produced by erosion for cropland is in the region of 17 471 000 tonnes annually and of this 30% will find its way into the river systems. Non-cropland, which includes rangelands (open grasslands that are used for grazing), forest and roads, will produce 1 646 000 tonnes annually and 11% of this will be deposited into the region's watercourses (USDA, 1979). These rates are dependent on regional topography, which is typically rolling hill country. Land in this country has been classified into Land Capability Classes where percent slope and mean rate of soil erosion has been modelled and it is this erosion model that has led to changes in agronomic practice and reductions in soil erosion (Figure 2.5).

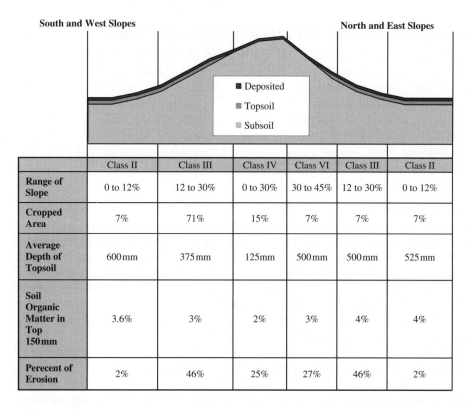

	Class II	Class III	Class IV	Class VI	Class III	Class II
Range of Slope	0 to 12%	12 to 30%	0 to 30%	30 to 45%	12 to 30%	0 to 12%
Cropped Area	7%	71%	15%	7%	7%	7%
Average Depth of Topsoil	600 mm	375 mm	125 mm	500 mm	500 mm	525 mm
Soil Organic Matter in Top 150 mm	3.6%	3%	2%	3%	4%	4%
Perecent of Erosion	2%	46%	25%	27%	46%	2%

Figure 2.5 Land classification and model slope profile of the cropped land in the Palouse Region

Agronomic Practice and Soil Conservation

The prevailing environment dictates the agronomy of the Pacific North West and USDA has developed descriptive criteria as a guide to cropping east of the Cascade Mountains

(Table 2.12). When growers follow certain rotations with a fallow in high precipitation areas soil erosion rates are typically high (Table 2.12). A wheat-pea rotation will lead to erosion rates of 20 tonnes/acre/annum (49 tonnes/ha/annum) because this rotation requires repeated tillage operations such as ploughing and subsequent harrowing, resulting in finer soil particles at the soil surface, which can then readily erode. When this crop rotation is compared to an annual grain and green fallow (such as wheat) there is a level of surface residue that provides a protective surface that impedes surface water run-off. The associated root channels, from the green fallow, aid the percolation of winter precipitation through the soil profile and thus reduce sheet erosion events. When this is compared with a winter fallow that has had the previous stubble burnt off and ploughed, the soils cannot hold the winter rains and rapidly become water logged and consequently run-off and erosion occurs. The advent of new seeding technology and plant breeding has enabled researchers, in conjunction with growers, to develop new tillage practice and combinations of crop rotations for the agronomic zones in the Pacific Northwest.

Tillage Practice, Crop Rotations and Contour Cropping

Adopting minimum tillage or no till practice can significantly reduce soil erosion, but the successful application of these approaches, known as conservation tillage, requires a considerable shift in crop rotations. To use these conservation tillage practices it is essential to control the disease and weed pressure, which is partly achieved by crop rotations and judicious applications of pesticides, in particular herbicide. The design of each new crop rotation has to consider the climatic zones occupied, and, to simplify the concept the following crop rotations are based on precipitation zones documented for the Palouse region which are for low, intermediate or high rainfall.

Low Rainfall

A typical rotation in the low rainfall region is based on spring cropping, as the autumn rains tend to arrive late and thus grass weeds such as downy brome (*Bromus tectorum*) cannot be sprayed out before the sowing of any winter cereal. Thus cropping is restricted to spring cereals into residual stubble, avoiding non-residue fallow and subsequent erosion. To achieve an economic yield growers need to control the disease and weed thresholds in the residue-fallow and the following rotations give reasonable cultural weed management that is boosted by applications of broad-spectrum herbicide (such as Roundup, which has been banned in Europe since 2007) before spring seeding.

A model rotation in the low rainfall zone of the Palouse region is spring wheat > spring wheat > spring barley > spring broadleaf (such as oil seed rape, mustard and safflower). The spring wheats provide the cash crop and the spring barley provides a break from soil borne diseases associated with spring wheats and the incorporation of broadleaved crops in the fourth year further aids disease management by reducing species-specific pathogens. In a cool moist spring flax (*Linum spp.*) and/or oats could be substituted for one of the spring wheats (Mallory *et al.*, 2000a).

Best practice using these rotations is the retention of a tall residue and typically tall stem varieties of cereals are grown and the header of the combine harvester is set to leave high stubbles. High stubbles trap winter snows and impede surface winds and thus prevent excessive thaw and wind erosion. Another advantage of high stubbles is the stems reduce wind-induced evaporation from the dry cold winter winds and thus soil moisture retention is increased, aiding establishment of spring seeding. Residue management must take into account the spreading of surface straw and chaff and this is achieved through a combination of choppers and chaff spreaders on the combine. Distributing the chaff aids uniform germination of weed seeds and helps in uniform application of pre-drilling herbicide application and a good level of weed control.

Intermediate Rainfall

The intermediate rainfall region allows more flexibility in the design of crop rotations and in this region it is possible to introduce a winter cereal into the rotation.

A typical three-year rotation is winter wheat > spring barley > legume or fallow. Moisture and weed thresholds are important in deciding on either a legume or a fallow. The rationale behind the three-course rotation is that winter wheat can be raised every third year on the same plot with acceptable control of weeds and disease, this has been generally acceptable but some grasses are problematic where a two-year break from a winter crop does not give acceptable control (Mallory et al., 2000b). The selection of spring barley as the second crop is on economic grounds, spring barley is less expensive to grow than spring wheat, and barley performs well following direct seeding into wheat stubble. In the intermediate rainfall zone spring stubbles can lay wet and cold and direct seeded barley will tolerate these conditions (Mallory et al., 2000b). The use of chemical fallow is to avoid conventional fallow on the slopes and hills and although chemical fallow can lead to moisture problems in the following autumn sown winter wheat, the advantages over conventional fallow with respect to soil retention on slopes clearly outweighs the slightly negative implications of a dry seedbed. Deep placement of winter wheats after a chemical fallow helps to avoid droughty seedbeds.

High Rainfall

There are several crop rotations applicable to the high rainfall zone and one system that uses a combination of grasses, cereals, legumes and oil seed rape is based on a five-year rotation. Starting winter wheat > spring cereal > spring legume > winter wheat > bluegrass, this rotation gives 40% winter wheat, which again is the cash crop. The development of new pea varieties, afila-type, which have interlocking tendrils gives a better residue in the winter compared to the traditional varieties. The absence of interlocking tendrils produced poor stubble that could be reduced to dust by high winds, leaving the soil and the young winter wheat plants exposed to damaging high winds.

Bluegrass (Poa spp.) requires a regular burn to rejuvenate the sward and restrictions on bluegrass burning in many areas in the Pacific Northwest have seen some growers remove bluegrass from their rotation. However, bluegrass is an excellent crop for managing soil erosion and fertility and some growers have developed a residue management system that encourages plant rejuvenation, where burning cannot be used. A reasonable strategy is once the seed has been harvested the grass should be baled, the stubble flailed and then deep harrowed two to three times in the autumn to spread the residue. This system works well on moderate slopes but cannot be used on steep slopes.

Using a five-course rotation has reduced many carry-over pathogens and improved yields but simultaneously resulted in heavy loads of straw and dense stubble, which has considerable implications on tillage strategy. In these scenarios chaff spreaders and straw choppers on the combine help but autumn tillage is required, which could be a combination of disc ploughing after the harvest followed by chisel ploughing in the autumn. This is essential in this region, as a spring chisel would result in lost seedbed moisture. Using this system all growers need to do to prepare for spring seeding is to spray the green-bridge and smooth the ground. Using a chisel in autumn leaves the ground rough, therefore water infiltration is not impeded and the substantial residue cover provides good protection from any surface erosion, (Mallory *et al.*, 2000c).

The crop rotations described provide a small subset of the rotations used by growers in the Palouse region. They have been designed to fit into a given farm system and maintain economic viability of the regions growers and simultaneously reduce the impact of continuous arable agriculture on soil health. This last point is paramount as a healthy soil is the essential for the economy of the region and ecological integrity of the watersheds and river systems. Using these new rotations growers can adopt a no-till or minimum tillage philosophy, increase fallow residue (green-bridge) and consequently maintain soil health.

Growers responding to research surveys have reported increased soil fauna, including earthworms, which secrete mucilage that helps to bind soil particles together and improve soil structure (Mallory *et al.*, 2000d). Other noticeable benefits include less soil disturbance on slopes using a direct drill, and consequently less soil erosion, improved root runs where stubble is present and less surface compaction leading to increased percolation of surface water during heavy rainfall. The over winter stubble also helps to maintain soil moisture in the upper soil horizon (300 mm) and where there are heavy cereal stubbles, surface moisture can be significantly higher, typically as much as 62.5 mm compared to fields with poor or no stubble where values average out at 45 mm (Mallory *et al.*, 2000e).

The key to the successful economic use of conservation tillage is the management of the crop residue (green-bridge), in particular, the weed management. Soil-borne crop pests and fungal disease can be kept at a tolerable economic threshold with successful crop rotations and, in some cases, applications of fungicide and insecticide. However, weed management is more complicated, in particular controlling grass weeds in cereal rotations, particularly where a cereal is used more than twice in a successive rotation. Broadleaf weeds can be managed by applications of a non-selective herbicide to the green-bridge two to three weeks before seeding, but grass weeds are more complicated. Their life-history strategy is very similar to the cereals, indeed both grasses and cereals are of the Poaceae family. Grasses are adapted to grow in the arable environment; in particular, they have an affinity for the cereal field. They can germinate and persist in the cereal understorey and therefore build up significant populations of individual propagules that remain at the soil surface in no-till practice. Thus controlling grass weeds in a cereal-dominated no-till environment relies heavily on herbicide application and the range of active ingredients is limited, giving rise to strong selection pressure on the resident population of grasses and eventually resistance to the limited range of active ingredients will develop.

A major issue associated with the heavy applications of pesticides, in particular herbicides, across a large agricultural region such as the Palouse, is the risk of these compounds entering watercourses. Even with the use of no-till and minimum tillage soils do still erode and many pesticide compounds are bound to soil particles and eventually find their way into surface and ground waters. The United States Geological Service (USGS) has demonstrated that six pesticides, one breakdown product and five volatile organic compounds (VOCs) were detected in ground water in the Palouse region and 29 pesticides and their breakdown products were detected in the surface waters. Sampling for these compounds took place three times a month between March and May in 1993/1994 and six times a month between November 1994 and February 1995 (USGS, 1997). Of these compounds, herbicides are the most prominent in surface waters and one herbicide compound exceeded the freshwater chronic criteria for the protection of aquatic life, and the concentration of two other herbicide compounds was approaching this threshold (USGS, 1997). Of all the pesticides detected, none were found in the groundwater that exceeded the drinking water standards, which was good news as this provides the only source of drinking water for this region (USGS, 1997).

The above discussion illustrates that even when ecologist and agriculturist try to develop new protocols for reducing environmental damage due to historical practice there are inevitably always some detrimental side effects. In this case the management of the green-bridge may lead to an increase in surface water contamination by pesticides. This could be partially alleviated by introducing riparian strips (Chapter 5) along the region's watercourses; the Conservation Reserve Programme (CRP) could support such activity.

The impact of conservation tillage on soil erosion is quite significant and across the three rainfall regions conservation tillage has reduced erosion rates by 35% (USDA, 1978). This equal reduction in soil erosion across the three rainfall regions is the result of changes in crop performance (high crop growth rates in moist areas and therefore more stubble) compared to low crop growth rates in low precipitation areas but generally less rain and reduced slopes. When conservation tillage is applied in conjunction with strip-cropping and divided slope farming soil erosion rates are further reduced (Table 2.13). Terracing has been excluded because it has limited impact (on average a 10% reduction in soil erosion) and is unfavourable to growers due to costs and unsightly development of weedy flora. This last point is worthy of mention. If growers could develop a vision of the 'unsightly' flora similar to that taken by organic cereal farmers, where such large strips of vegetation are seen as a refuge for insect predators which will spill out into the crop and aid in pest control.

Table 2.13 The impact of strip cropping and divided slope farming on soil erosion in the Palouse region

Conservation practice	% reduction in soil erosion		
	Precipitation zone		
	Low (300–375 mm)	Intermediate (375–450 mm)	High (450 mm+)
Strip-cropping	28	15	24
Divided slope farming	28	15	24

Adapted from USDA (1978).

Another development of conservation tillage is the placement of fertiliser below the seedbed and this has helped to reduce nutrient run-off and improved crop yields as the fertiliser is placed close to the developing root zone. When this practice has been combined with widespread conservation tillage the nutrient enrichment of watercourse from non-point source has declined over time (Wise *et al.*, 2007).

On balance the STEEP Programme has made significant progress in reducing soil erosion in the Pacific Northwest and since its inception the uptake of conservation practice has been significant (Table 2.14). This confirms that the methods developed by STEEP are both economically attractive and sensitive to environmental issues. However, there are still significant problems and in preparing this case study one observation is paramount; that reporting on wildlife conservation is peripheral and not integrated into the philosophy of the region's farming sector. This is a shame, as the total integration of farming and wildlife conservation would simply require a few additional steps as detailed below:

- Report positive impacts of conservation tillage on soil microfauna.
- Report positive impacts of conservation tillage on aquatic life.
- Integrate bluegrass strips and CRP strips into crop rotations to maintain invertebrate diversity.
- Introduce riparian strips and tree planting as an integrated farm system.
- Introduce and integrate conservation strips for game cover and barriers to soil erosion.
- Develop a five- to six-year rotational grass ley to improve soil fertility and quality.
- Integrate conservation strips, CRP strips, riparian planting and grass leys to improve landscape heterogeneity.
- Disseminate the above points in technical bulletins and at farm demonstrations.

Table 2.14 Summary of achievements of the STEEP programme

	Hectares under erosion control		Predicted average annual reduction in erosion[a]	
Erosion control practice	1979	1994	tonnes/ha	Total tonnes
No-till	242	22 672	22	500 000
Conservation reserve programme	2591	24 534	12.35	270 000
Strip-cropping and divided slopes	0	96 761	–	–
Grass water-ways[a,b]	24	470	2.47	1500
Planting trees and shrubs	0	1486	24.7	37 000
Conservation tillage	0	32 793	19.76	650 000
Totals	2857	178 716	–	1 458 500

[a]Predictions from universal soil loss equation (Wischmeier and Smith, 1978). For grass waterways based on gross erosion prediction method (Renard *et al.*, 1997).
[b]Linear metres.
Adapted from James and Dennis (1998).

Case Study 2.3: Cholderton Estate

Cholderton Estate is a 1000-ha estate straddling the counties of Wiltshire and Hampshire, comprised of a mixed organic farm, woodlands, a private water company and a light industrial unit. The farm consists of 202 ha (500 acres) of arable crops, 162 ha (400 acres) of woodland, two dairy herds totalling 270 dairy cows, consisting of black and white Holstein and dairy shorthorns. The farm also runs a beef suckler herd, an extensive flock of Hampshire Down sheep (250 ewes), which have been on the farm for over 110 years, and a pig unit including rare breeds such as Tamworths.

Another important component of the estate's livestock is a Cleveland Bay stud. These horses have been bred on the estate for over 120 years and the estate also cares for 80 polo ponies throughout the winter months.

Cholderton Estate is owned by Henry and Felicity Edmunds, Henry started managing the business in 1975, and has an extraordinary vision for the estate that does not align with modern industrial scale farming. Henry's view of the business and consequently the local landscape is one that dates back to the pre-industrial era, where food production systems were based on a low input philosophy and subsequently the regional landscape diversity is enhanced by farming practice that values crop rotations, hedgerows, semi-natural grasslands and healthy water courses. The estate is recognised by Hampshire County Council, which has designated Cholderton as a Site of Importance for Nature Conservation (SINC[1]), indeed the council launched its biodiversity programme at Cholderton Estate.

Landscape Character

Cholderton lies on the far edge of the Hampshire Downs and is adjacent to the Wiltshire Downs, with parts of the estate forming a boundary to the southern end of Salisbury Plain. The area is dominated by large-scale arable agriculture, with cereals forming the major crop rotation. The Hampshire Downs form the highest points of the broad belt of chalk in Hampshire, retaining a rural character of rolling hills, country lanes and small settlements.

The landscape types are open arable, chalk and clay, clay plateau and downland and hanger scarps, which support a diverse array of habitats; ancient semi-natural woodland, parkland and pasture woodland, hedgerows, lowland calcareous grassland, arable, open standing water and chalk streams. Many of these habitats occur on Cholderton Estate and are maintained in good condition by the farming philosophy adopted by Henry.

Cholderton lies adjacent to three designated areas: Porton Down, Salisbury Plain and East Hampshire Hangers SSSIs and two further SSSIs, and two AONBs. Taken as a whole, the landscape character of the Hampshire Downs is generally elevated chalk with sweeping contours varying from gently domed or undulating clay plateaux to steeply rolling and domed hills with dry valleys and combs. Superimposed onto this matrix are a number of dramatic escarpments forming prominent skylines of woodland or exposed hilltops. As the region is dominated by arable crop production systems the field systems in the region are large and often very open with few hedgerows and thus, providing few if little

---

[1] A site of county wildlife importance.

opportunities for farmland wildlife. It is this observation that puts Cholderton Estate on the map. Due to the forward and unconventional philosophy adopted by Henry, towards large-scale farming, Cholderton has retained many of its hedgerows, pockets of woodland and numerous areas of semi-natural open calcareous grasslands (Map 2.3). This has been achieved by sound economic decisions in the way the farm business is structured, which has culminated in a unique and unequalled matrix of habitats, supporting both a viable business and numerous species of plants and animals, including many rare and threatened species.

Crop Systems

The key to any organic system is soil fertility and unlike conventional farming systems, where fertility is artificially maintained by the use of inorganic fertilisers, the organic system relies on crop rotations and FYM. The system adopted by Henry is based on an historic fodder crop, sainfoin (*Onobrychis viciifolia*) a plant from the pea, leguminoseae, family. A Mr Hartlib introduced sainfoin to the UK as a forage crop in 1652, when he shipped a sack of SAINT FOINE from Calais, noting that the plant thrived on poor dry chalky soils (Maybey, 1996 and www.cotswoldseeds.com – accessed 6 December 2009).

Sainfoin

In the UK the native sainfoin occurs on the dry chalk grasslands of southern England (www.naturalengland.org.uk – accessed 6 December 2009) and is somewhat smaller in growth form and habit compared to the more commonly observed, improved commercial variety, which persists as a relic of historic farming systems across southern England. The commercial sainfoin (Figure 2.6) is a medium sized erect to semi erect, pubescent plant with hollow stems, reaching a final height of 15–90 cm. Leaves are pinnate with 6–14 pairs of oblong, occasionally linear leaflets 15–80 mm long. Flowers are pink with purple veins, 10–14 mm long in long-stalked axillary racemes with a toothed calyx. Pods are small, 5–8 mm with a toothed margin (Blamey and Grey-Wilson, 1989). Sainfoin relies on cross-pollination for flower set and reproduction; Kropacova (1969) demonstrated that seed production from plants exposed to pollinators produced 179 g/m^2 compared to 9.75 g/m^2 of caged plants, and commented that the pollinators were reported to be honeybees. The recommendation for honeybee colonies is two to three colonies per hectare and on our visit to Cholderton the aerial environment above and around the sainfoin flowers was humming with honeybees, compared to the adjacent white clover, which appeared, by comparison, to be empty of honeybees (Kropacova, 1969).

Like all members of the pea family, Sainfoin is able to fix atmospheric nitrogen through the rhizobial nodules characteristic of plants from the pea family. The root system of sainfoin consists of a deep taproot with numerous side branches and many fine lateral roots bearing most of the rhizobial nodules, the presence of the nitrogen fixing nodules negates the use of inorganic nitrogen. Some conventional growers will add 40 kg of nitrogen per hectare at sowing to aid establishment (www.cotswoldseeds.com – accessed 6 December 2009). The long taproot is an adaptation to its native habitat of dry calcareous soils

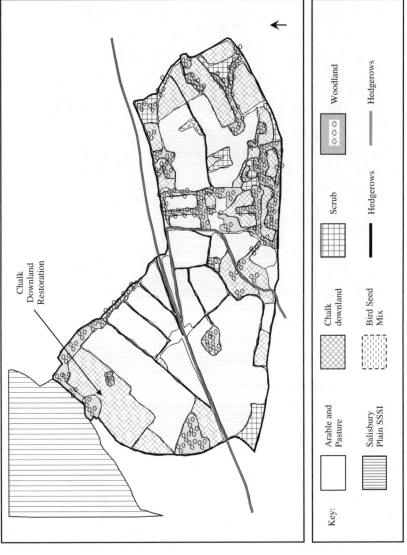

Map 2.3   Cholderton Estate, illustrating a proportion of the estate and local landscape matrix

Key:

Arable and Pasture

Salisbury Plain SSSI

Chalk downland

Bird Seed Mix

Scrub

Hedgerows

Hedgerows

Woodland

Hedgerows

Chalk Downland Restoration

Figure 2.6   Sainfoin. (Photo: Stephen Burchett)

and partially explains its persistence in these droughty soils and subsequently an ideal choice for the farming system at Cholderton where sainfoin is used as a component of the arable rotation and as a forage or hay crop. Sainfoin is intolerant of prolonged wet spells and soils in which the water table is high.

The traditional and local strain of sainfoin to Cholderton is the Hampshire Common strain, which was used as a hay crop and grazed by ewes and lambs. At Cholderton sainfoin is still used as a hay crop and is part of the grazing regime for the estate's livestock. It forms an integral component of the crop rotations improving soil fertility. The choice of sainfoin as a fodder crop makes excellent sense in terms of animal health, sainfoin is very palatable and reportedly is non-bloating (Edmunds: personal communication, 2008) and with 20% protein content animals grazed on sainfoin perform considerably better than animals grazed on grass alone (www.cotswoldseeds.com – accessed 6 December 2009). Sainfoin has natural anthelmintic compounds and consequently helps to control parasitic worms in organically raised livestock, where Ivermectin based worming compounds are not permitted. As a conserved forage, productivity ranges from 2.5 to 5 tonnes/ha, considerably less than conventional grass based systems, but the reduced inputs (no nitrogen and phosphates) offsets production costs and, when combined with reduced costs associated with prophylactic animal drugs and increased animal welfare, sainfoin is a clear choice for organic systems located on light dry soils.

Crop rotations are fundamental at Cholderton and the estate relies on the following rotation to build soil fertility:

- Grass ley > legume > cereal > oilseed rape/kale > cereal whole crop > grass ley.
- Legume > oats > cereal whole crop > oilseed rape/kale > grass ley/legume.

The two legumes used, sainfoin and hairy vetch, fix atmospheric nitrogen and the grass ley builds up soil organic matter and soil nitrogen from the clover component. The break in the cereal crop with the broad leaf crops is part of the weed and disease control strategy, where grass weeds can be controlled and out competed by the heavy foliage of the broad leaf crop. Grass weeds become problematic in continuous cereals as they occupy a similar niche as the cereal crop. Initially in cereal crops the seed bed appears very weedy but the broad leaf weeds soon become suppressed by the developing canopy of the cereal crop and modern combine harvesters are very efficient at separating cereal grains and weed seeds during the harvest, helping to reduce the return of weed seeds to the seed bed.

Another important element in the control of weed seeds in cereal crops is establishing a dense canopy that is achieved by manipulating seed rates. This can be problematic as low seed rates result in low populations of cereal plants (100–150 plants/m^2) and consequently poor competitive ability against the weed population. However, low seed rates compensate by producing more ears per plant (average 2) and more grains per ear (Finch, Samuel and Lane, 2002) but to reduce the competitiveness of weeds higher populations of cereal plants are required but thick dense crops (400+ plants/m^2) are susceptible to lodging and disease problems (Finch, Samuel and Lane, 2002). The seed rate for oats – the staple cereal at Cholderton – adopted by Henry is quite high at 201 kg ha giving a final plant population between 472 and 508 plants/m^2 depending on individual field conditions. This gives excellent weed control (Figure 2.7) and, as there is no artificial nitrogen applied to

Figure 2.7 Organic oats. Note dense plant populations and closed canopy with minimal weed populations. (Photo: Stephen Burchett)

the crop, the plant growth is not lax and sappy but firm and able to resist lodging and foliar disease. The droughty soils, prevailing benign environment and rigorous application of crop rotations also help to suppress foliar pathogens. This example illustrates how essential it is to understand the subtle nuances of agronomy and that by following sound application of this knowledge the estate is able to develop an integrated farm system, which builds on a balance of mixed cropping and livestock production. This philosophy reduces the reliance on inorganic fertiliser and the prophylactic use of damaging synthetic pesticides, maintains habitat diversity within the farm and consequently improves opportunities for species conservation.

Conservation Work

Conservation work at Cholderton is as diverse as the farm business. The rigorous applications of crop rotations has increased local heterogeneity and consequently improved species diversity (Benton, Vickery and Wilson, 2003). Good examples of conservation outcomes are commonly observed across the estate, for example standing at a gateway to a field sown to vetch the aerial environment just above the crop was alive with swallows (*Hirundo rustica*) and swifts (*Apus apus*) foraging a diverse array of invertebrate fauna. Indeed, vetch plants adjacent to the gate were buzzing with numerous hymenoptera and diptera species and this was further enhanced by the wide margins of tall semi-natural grassland. Another example of species diversity in relation to local heterogeneity was the appearance of two hares (*Lepus europaeus*) (a BAP species) running out of a fertility building crop, this time a field split and sown with sainfoin and a grass ley, where the grass ley had just been cut for hay.

This is not the only conservation effort occurring at Cholderton. A significant amount of staff-hours each year is devoted to scrub management in a number of semi-natural grassland areas across the estate (Figure 2.8), with the aim of maintaining open grassland

Figure 2.8 A typical grassland and scrub landscape at Cholderton Estate. Regular scrub clearance, every 18–24 months maintains the ratio of grassland to scrub. Grazing maintains sward structure. (Photo: Stephen Burchett)

Figure 2.9   Bastard Toadflax (*Thesium humifusum*). (Photo: Stephen Burchett)

habitats for important floral such as meadow clary (*Salvia pratensis*), bastard toadflax (*Thesium humifusum*) (Figure 2.9) and greater knapweed (*Centaurea scabiosa*), where the first two species have a very restricted distribution in the UK. Many of these grassland forbs are essential nectar resource plants for a number of invertebrate species and attract numerous invertebrates including hymenoptera; lepidoptera and diptera and consequently these grasslands are alive with invertebrate fauna (Figure 2.10). Chalk downland grassland is also being restored (Map 2.3), supported by agri-environment payments currently under the CSS, which will eventually be replaced by the new agri-environment scheme known as Higher Level Scheme (HLS) (Chapter 1). The challenge to complete restoration is excessive grazing of newly sown seed or planted plugs by rabbits from the adjacent warren. Henry has developed several strategies to enhance the establishment of new plants, which include erecting exclusion cages around single plants and designing a sowing guard, where wire mesh is stretched across a softwood frame and laid down on top of freshly sown patches (Figure 2.11). These sowing guards prevent rabbits from scratching out germinating and young plants. Again this is time consuming but the results are encouraging and floristic diversity on the downland sites under restoration is high.

Figure 2.10 Marbled white (*Melanargia galathea*). (Photo: Stephen Burchett)

Figure 2.11 Novel planting frame used to protect new sowings and or plug planting of calcareous grassland species from rabbit damage. (Photo: Sarah Burchett)

In conclusion, Cholderton Estate is an excellent example of the marriage between farming and conservation. The underlying organic philosophy clearly enhances the structural diversity across the estate, which has improved resource opportunities for local flora and fauna. The Cholderton Estate illustrates how good agricultural practice has been aligned with knowledge of the ecological requirements of local flora and fauna, which has then been applied across the estate in the form of habitat management. The focus on habitat management is an important element of the business, contributing to the commercial success as well as improved opportunities for local wildlife. This approach can also be applied to conventional farms and the Elveden Estate (above) has adopted a very similar philosophy of farming practice and habitat management.

Case Study 2.4: Blueberry Hill Farm, Maryland, US

We met Michael and Krissie James selling their fruit and vegetables at a farmers market in Shepherdstown, West Virginia; a young, enthusiastic couple with a passion for the ethos that drives people to choose organic farming over conventional farming. What was immediately obvious was the high quality of their produce; these were people who had to work extra hard to overcome the problems associated with organic farming and still maintain a high standard of produce to take to market. We were impressed enough that we asked them if they minded us coming for a visit to their farm, just across the state line into Clear Spring, Maryland the next day.

We arrived at Blueberry Hill Farm to find our hosts hard at work in their polytunnels where an array of handmade benches and raised beds supported a wide variety of young plants from tomatoes to carrots that were companion planted with onions. Companion planting is one of the key elements for pest control in organic farming. Outside, the practice of companion planting is perpetuated where brassicas and onions grow well together, (Figure 2.12) below. At this farm a number of species of caterpillar collectively known as cabbage worm can be a

Figure 2.12   Companion planting and trickle irrigation. (Photo: Sarah Burchett)

problem, the companion planting reduces this considerably, however a persistent infestation may warrant the use of Bt (*Bacillus thuringiensis*) to control these pests organically.

Blueberry Hill Farm is a tiny smallholding, the smallest of our case studies, indicating just what can be achieved in relatively little space. The main site is just 2.5 acres, (slightly over a hectare), with a further 5 acres (~2 ha) of mixed broadleaved woodland, owned by Michael's father, where approximately 800 blueberry bushes (*Vaccinium spp.*) are grown in clearings. Blueberries make a useful crop where pH is typically low as they favour a soil pH of 4.5–5.5. Bees pollinate the flowers, and the James' have their own hives on the site. Blueberries are not only delicious but are extremely beneficial to health, high in vitamins, especially vitamin C, they also have one of the highest levels of antioxidants of any fruit.

These woodland clearings were one big medicine cupboard as around the fringes of the clearings, morels (*Morchella spp.*) and shiitake mushrooms (*Lentinula edodes*) were being cultivated in deadwood piles (Figure 2.13). Though morels have been eaten for a very long time in the US, shiitake mushrooms are relatively new to the North American diet; they have however been cultivated in East Asia for centuries. The shiitakes are known for their high levels of beta-glucans which function as immune modulators. Making use of the mushrooms' saprophytic nature, spores are used to inoculate cut wood, these sites are then sealed up with wax plugs, (old deadwood is unsuitable as it may contain spores of other fungi). These are then left to their own devices and the fruiting bodies can be harvested periodically when ready.

Figure 2.13 Shiitake mushroom spores 'seeded' in logs and plugged with wax. (Photo: Sarah Burchett)

The James' use most of their own seed now that they have become established, but do link up with a seed exchange programme when necessary. On their plots they have a couple of mulberry trees *(Morus spp.)* and an apple orchard with raspberries, gooseberries and currants grown between the rows. For row crops they have pea trellises (fenced off against rabbit foraging), a variety of vegetables, salad crops and flowers for the cut flower market, plus a variety of green manure crops to help increase organic matter in the soil. Indoors mushrooms are grown, and the mushroom compost is also used for fertilisation. A number of greenhouses, polytunnels and cold-frames are used on the farm. There is also a plot for herbs, both for direct sale at market and Krissie thinks they may produce natural toiletries in the future. There were a few chickens around the farm when we visited, and Michael was in the process of renovating an old chicken coup with a view to increasing fowl and egg production. Irrigation is carried out by a trickle system (Figure 2.12), and plans were afoot to put water butts in place ready for the autumn rains.

Temperatures in Maryland vary from about 30 °C in the summer to below freezing in the winter. Greenhouse temperatures are maintained by heating from the house in the winter, and by cooling using an old cooling unit from MacDonald's in the summer. Even so, winter can be a lean time for farmers in temperate regions and diversification is a must, in this case pottery production added to the range of goods to go to market. We asked them about self-sufficiency, they said that the farm was second generation and the house paid for. If they had to pay a mortgage or rent they did not believe that this would be possible, however without that extra cost, they may well be self-sufficient. They were keeping ongoing records to answer that question for themselves.

The whole ethos of organic farming is important to the James family, from a health viewpoint, they are keen to raise their young daughter in a clean, chemical-free environment and they have high quality products with an eager clientele in nearby towns and in Washington where they regularly travel to market. However, the main driving force was the desire to encourage wildlife into the holding. Wild turkey (*Melegris gallopavo*) numbers are on the increase, and Michael had put up extra posts for visiting birdlife including thrushes, bluebirds and warblers (a keen interest of his), which were clearly evident during our visit, where many birds flew in from the distant woodlands not stopping en route to visit the conventional farmland surrounding Blueberry Hill. Butterflies and other invertebrates also thrive in this benign environment, and we found the James's keen to learn more about hedgerows and the benefits to wildlife and to farming practice that can be gained from planting hedges (a strangely English practice). Overall, an inspiring tale of young folk with a passion for benevolent farming practice and for wildlife conservation.

2.15 Chapter summary

The underlying theme throughout this chapter has been the establishment of knowledge and understanding of contemporary agriculture with an emphasis on mixed farm systems. Using a model crop system, winter wheat, students should be able to appreciate the exacting requirements that modern food production systems impose on farmers. The conventional model can be compared to the organic approach where success relies on the management of soil fertility. In both conventional and organic systems students should appreciate that natural habitats are perturbed and hence mitigation of agricultural operations on the adjacent environment forms the basis of good agricultural practice. This concept has been reinforced by four case studies, which illustrate a range of approaches to farming and wildlife conservation.

Starting with Down Farm in south Devon, where the focus is on integrated farm management and wildlife conservation is supported by government grants, to a tri-state programme in the USA where the key environmental driver is soil conservation, again supported by government funding and advice. Cholderton Estate illustrates how a modern large-scale organic farm can adapt its business model to incorporate a mixed farm system into a landscape of chalk downland grassland, adding a complex mix of habitats into a landscape that is dominated by modern conventional cereal production. The estate has developed a philosophy of soil conservation and has designated large areas of land to semi-natural grassland and consequently conserving an array of flora and fauna. Again the agronomy is based on good agricultural practice that results in the integration of production and conservation. The final case study demonstrates how a very small farm can contribute to the conservation of biodiversity by adopting sound agronomic principles into their small production system.

Throughout the development of farming, wildlife has evolved to cope in an ever-changing landscape and until the 1960s this change has been at a relatively slow pace but the advances in agricultural technology meant that the pace of change in the rural landscape has been colossal, leading to significant detrimental impacts on biodiversity. The case studies presented in this chapter demonstrate that small adjustments in farming practice, often just adopting good agricultural practice, can have positive impacts on local wildlife, but to enhance these gains, in particular where species can be classified as specialist species, it is essential to understand both the biological and ecological requirements of the target species and to match farming practice to these requirements. This may require the conservation biologist to consider how a target species may have coevolved with farming, such as the rare arable flora or the cirl bunting and consequently design a conservation strategy that incorporates this knowledge into farming practice. However, this may be problematical in terms of financial security for the farmer as this may mean adopting a production system with reduced yields and crop quality, in which case such conservation work could be supported by government grants or tax aid.

References

Bahn, P.G. (2006) *Atlas of World Archaeology*, Sandcastle Books, Worcester.

Benton, T.G., Vickery, J.A. and Wilson, J.D. (2003) Farmland biodiversity: is habitat heterogeneity the key? *Trends in Ecology and Evolution*, **18** (4), 182–188.

Blamey, M. and Grey-Wilson, C. (1989) *The Illustrated Flora of Britain and Northern Europe*, Hodder and Stoughton, London.

Clough, Y., Holzschuh, A., Gabriel, D. *et al.* (2007) Alpha and beta diversity of arthropods and plants in organically and conventionally managed wheat fields. *Journal of Ecology*, **44**, 804–812.

Cove, M. (2006) *Geology of the South Hams Coast. Coastline Rocks, Bigbury Bay to Start Bay, Devon*, Hedgerow Print, Crediton, Devon.

Defra (2000) *Fertiliser Recommendations for Agricultural and Horticultural Crops. RB209*, The Stationery Office, Norwich.

Douglas, C.L., Chevalier, P.M., Klepper, B., *et al.* (1999) Conservation cropping systems and their management, in *Conservation Farming in the United States. The Methods and Accomplishments of the STEEP Program* (eds E.L. Michalson, R.I. Papendick and J.E. Carlson), CRC Press. Pp. 73–77.

Finch, H.J.S., Samuel, A.M. and Lane, G.P.F. (2002) *Lockhart and Wiesmans Crop Husbandry: Including Grassland*, 8th edn, Woodhead Publishing Limited, Cambridge.

Gibbson, R.H., Pearce, S., Morris, R.J. *et al.* (2007a) Plant diversity and land use under organic and conventional agriculture: a whole farm approach. *Journal of Applied Ecology*, **44**, 792–803.

Gibbson, R.H., Pearce, S., Morris, R.J. *et al.* (2007b) Plant diversity and land use under organic and conventional agriculture: a whole farm approach. *Journal of Applied Ecology*, **44**, 425–432.

Grime, J.P. (1974) Vegetation classification by reference to strategies. *Nature*, **250**, 26–31.

Hesketh, S. (2006) *Devon's Geology: An Introduction*, Bossiney Books Launceston, Cornwall.

HGCA (2004) *Determining Eyespot Risk in Winter Barley*. Topic Sheet No 80. Available at http://www.hgca.com (accessed 5 December 2009).

Jellings, A. and Fuller, M. (2003) *The Agricultural Note Book*, John Wiley & Sons, Inc.

Kropacova, S. (1969) The relationship of honey bee to Sainfoin (Onobrychis sativa). 22nd International Apicola Congress Proceedings, Munich, pp. 467–477.

Lampkin, N. (2001) *Organic Farming*, Old Pond, Ipswich. Pp. 120–160.

Mallory, E.B., Fiez, T., Veseth, R.J. *et al.* (2000a) *Direct Seeding in the Inland Northwest. Ron Jirava Farm Case Study*, Pacific Northwest Publications.

Mallory, E.B., Fiez, T., Veseth, R.J. *et al.* (2000b) *Direct Seeding in the Inland Northwest. Ensley Farm Case Study*, Pacific Northwest Publications.

Mallory, E.B., Fiez, T., Veseth, R.J. *et al.* (2000c) *Direct Seeding in the Inland Northwest. Schultheis Farm Case Study*, Pacific Northwest Publications.

Mallory, E.B., Fiez, T., Veseth, R.J. *et al.* (2000d) *Direct Seeding in the Inland Northwest. Aechliman Farm Case Study*, Pacific Northwest Publications.

Mallory, E.B., Fiez, T., Veseth, R.J. *et al.* (2000e) *Direct Seeding in the Inland Northwest. Jepsen Farm Case Study*, Pacific Northwest Publications.

Maybey, R. (1996) *Flora Britannica*, Sinclair-Stevenson, London.

McCool, D.K. and Busacca, A.J. (1999). Measuring and modeling soil erosion and erosion damages, in *Conservation Farming in the United States: The Methods and Acomplishments of the STEEP Program*. Michalson, E.L., Papendick, R.I. and Carlson, J.E. Eds. CRC Press Boca Raton, Florida.

Papendick, R.I., McCool, D.K. and Krauss, H.A. (1983). Soil conservation: Pacific Northwest, in H.E. Dregne and W.O. Wills, Eds. *Dryland Agriculture, Agronomy Monograph 23, American Society of Agronomy, Crop Science Society of America, and Soil Science Society of America*, Madison, WI, pp. 273–290.

Philipps, L., Huxham, S.K., Briggs, S.R.S.R. *et al.* (2001) Rotations and nutrient management strategies, in *Organic Cereals and Pulses* (eds D. Younie, B.R. Taylor, J.P. Welsh and J.M. Wilkinson), Chalcombe Publications, Lincoln. Pp. 51–75.

Powell, M.L. and Michalson, E.L. (1986). An evaluation of best management practices on dryland farms in the upper snake river basin of Southeastern Idaho, *Idaho Agricultural Experimental Station, Bulletin 655*, University of Idaho, Moscow.

Rackham, O. (2006) *Woodlands*, Collins, London.

Renard, F.G., Foster, G.R., Weesies, G.A. *et al.* (1997) *Predicting Soil Erosion by Water: A Guide to Conservation Planning with Revised Universal Soil Loss Equation (RUSLE).* U.S. Department of Agriculture.

Stubbs, A.E. (1994) *1993 Terrestrial Invertebrate Survey for Prawle Point-Start Point SSSI, South Devon*, English Nature, Taunton and Oakhampton, English Nature Report 126.

USGS (1997) *Pesticides and Volatile Organic Compounds in Ground and Surface Water of the Palouse Subunit, Washington and Idaho.* USGS Fact Sheet 204–96. Available at http://wa.water.usgs.gov/pubs/fs/fs241-95/ (accessed 18 October 2009).

Wheat Growth Guide (2008) Available at http://www.hgca.com/cms_publications.output/ 2/2/Publications/Publication/The%20Wheat%20Growth%20Guide.mspx?fn=show &pubcon=4444 (accessed 12 December 2009).

Wilson, P. and King, M. (2003) *Arable Plants: A Field Guide*, English Nature and WildGuides, Old Basing Hampshire.

Wise, D.R., Rinella, F.A., Rinella, J.F. *et al.* (2007) *Nutrient and Suspended-sediment Transport and Trends in the Columbia River and Puget Sound Basins, 1993–2003.* U.S. Geological Survey Scientific Investigations Report 2007-5186.

Wischmeier, W.H.H. and Smith, D.D. (1978) *Predicting Rainfall Erosion Losses: A Guide to Conservation Planting.* U.S. Department of Agriculture Handbook No. 357.

Yearbook of Agriculture (1934). Eisenhower, M.S. Editor and Chew, A.P. Associate Editor. Dept of Agriculture, Washington, DC.

3 Grasslands

3.1 Introduction

The term grassland covers a wide range of geographical areas, from the Eurasian steppes to the African savannas, from the North American prairies to the South American pampas. They are, in fact, found on all the continents – apart from Antarctica – and are adapted to, and tolerant of, great variations in geology, climate, pH and salinity. As such, they form the habitat type for a vast array of ecosystems, overall covering approximately 25% of the world's land area. Africa is at the top of the scale with approximately 50% grassland. Loosely divided into temperate and tropical grasslands, these biomes are then sub-divided to incorporate numerous natural, semi-natural and cultivated ecosystems.

The family of grasses (*Poaceae*) has in excess of 10 000 known species (www.kew.org – accessed July 2009); this includes the cereals that are discussed in greater depth in Chapter 2. They cover vast areas of land around the world, but they do not live independently. Grasslands generally have a number of different grass species plus reeds, sedges and forbs – flowering plants that generally live in association with grasses such as the clovers (*Trifolium*) and the yarrows (*Achillea*). These complex plant associations are fundamental to their place in habitat management for both anthropogenic land use and wildlife conservation. Some of these grasslands are naturally occurring while most have developed under human influence, for example by deforestation and cultivation; either way, they represent the primary production for countless global habitats and their importance in conservation biology therefore cannot be overstated. In this chapter we have focused on just a few such ecosystems:

- Cultivated grassland for grazing domestic stock:
  — grass production;
  — breed selection;
- Making more of native grasslands:
  — grasslands of the United States;
  — grasslands of the British Isles;
- Grassland ecosystems around the world:
  — tropical;
  — temperate.

Introduction to Wildlife Conservation in Farming   Edited by Stephen Burchett and Sarah Burchett
© 2011 John Wiley & Sons, Ltd

3.2   Cultivated grasslands for grazing domestic stock

3.2.1   Grass production

We see grass wherever we go in the countryside, on sports grounds, along the roadside verges, popping up inconveniently between cracks in the pavement, plus a whole industry has built up around standardising that little patch of lawn in our gardens. We take grass so much for granted that it is hard for the non-farmer to understand the labour intensive and expensive difficulties associated with growing it commercially for pasture, haymaking and silage.

Domestic stock, especially dairy cattle require high quality, nutrient rich grasses; that little garden patch that's taken all year to get 'just so' wouldn't last for long if a hungry Friesian cow wandered on to it. Large-scale grass production for intensive grazing takes a lot of work.

Good soil structure is essential; it should be well aerated, not compacted. It should contain organic matter, which helps to keep nutrient levels up and the soil porous so that it is more likely to retain moisture. Compacted soils lead to roots not developing properly, water does not penetrate such soils well, and nutrient applications are likely to run straight off the top and possibly end up in watercourses. There are a number of methods at the farmers' disposal to improve soil structure including sub-soiling and applications of organic matter such as farmyard manure (FYM), maintenance thereafter is largely by good husbandry, such as preventing overgrazing, especially in very wet conditions when poaching is common around gateways and troughs (Figure 3.1). Such issues are discussed in greater detail with respect to soil erosion in Chapters 2 and 5.

Having spent time and money ensuring that the ground is ready, grass seed mixes are sown usually in the spring when light levels are increasing, the ground is warming up, but not baking and water is generally available. Seed mixes vary according to grazing stock requirements but usually contain a number of grassland species including perennial ryegrasses, fescues and forbs (Table 3.1). The seed mixes described in Table 3.1 are designed for modern commercial farming and are commonly called improved pastures and are typified by low species diversity, both in terms of grasses and forbs and do not truly represent the range of semi-natural grasslands observed in the UK. In the UK semi-natural grasslands are broadly divided into four types (Price, 2003):

- mesotrophic;
- calcareous or chalk;
- acidic;
- mires.

Grasses for grazing are typically nutrient-hungry and must be managed to ensure the production of good quality herbage and to minimise soil damage that leads to

Figure 3.1 Damage caused by 'poaching' around a water trough. (Photo: Stephen Burchett)

sward degradation and thus costly re-seeding. Table 3.2 shows a typical grassland year and in modern improved pastures dominated by Italian and perennial ryegrass (*Lolium perenne*) swards forage yields can be as much as 10–11 tonnes/ha dry matter for grazing and 13–14 tonnes/ha for silage production (DEFRA, 2000).

Good husbandry is not only more cost-effective for the farmer by removing the once common practice of prophylactic application of nutrients and pesticides, but also helps to maintain high standards of stock welfare and helps to prevent ongoing environmental impacts such as watercourse eutrophication and damage to marginal ecosystems flanking the farmland. To manage grassland effectively it is essential to understand the characteristics and history of grazing animals, as breed selection will directly impact on the characteristics of grassland sward and influence conservation outcomes.

3.3 Breed selection

One of the major issues with farming generally and significantly with raising livestock, is that, in the last few decades, we have moved more and more towards intensive farming for the high yields this can bring. This has been to the detriment of natural habitats and its associated wildlife. Furthermore it has been to the detriment of the smaller farmer trying to eke a living out of more marginal regions such as the uplands. This not only decreases biodiversity, but also decreases the diversity of products to be found in the supermarket.

Grazing can, however, be a positive force in the move to increase biodiversity on farmland, by specifically selecting stock animals that do not require high-nutrient

Table 3.1 Grass seed mixes

Production system	Seed mix (kg/ha)	Comments
Short term silage leys		
One year high yield	24.7 kg Fabio tetraploid Italian ryegrass and 9.9 kg Ligrande Italian ryegrass.	Produces a first cut silage from an autumn sowing, responds well to early nitrogen. High D values and high soluble carbohydrates, ensuring good fermentation.
Hybrid silage ley (3–4 year ley)	15 kg Aberecho tetraploid hybrid ryegrass, 15 kg Abereve tetraploid ryegrass and 5 kg Portrush perennial ryegrass.	Early growth and good persistence ensuring a 3–year ley. Comparable yields to Italian ryegrass but have greater sward density thus providing improved grazing.
Maximum D-value (4–5 year silage ley)	10 kg Aberglyn tetraploid perennial ryegrass, 15 kg Abereve tetraploid hybrid ryegrass, 5 kg Aberstar perennial ryegrass and 5 kg Portrush perennial ryegrass.	Produces high D-value silage during the third week of May (England). High yield obtained by the inclusion of hybrid ryegrass. This mix will also provide good late season grazing.
Two year red clover ley	15 kg Fabio tetraploid Italian ryegrass, 7 kg Ligrande Italian ryegrass and 7 kg Altaswede red clover.	The upright growth habit of Italian ryegrass enables good establishment of red clover. The clover produces good leaf growth and root development which enables the fixation of atmospheric nitrogen by the root nodules. Inclusion of the clover improves voluntary intake of protein.
Grazing mixtures		
Intensive dairy grazing (early growth 4–5 year ley)	11 kg Aberecho tetraploid hybrid ryegrass, 11 kg Aberstar perennial ryegrass, 7 kg Aberglyn tetraploid perennial ryegrass, 5 kg Ideal tetraploid ryegrass and 5 kg Portrush perennial ryegrass.	An early season grass ley that produces high quality medium term grazing. Good durability of palatable grasses.

Table 3.1 (*continued*)

Production system	Seed mix (kg/ha)	Comments
Dual purpose, cut and graze longterm ley	7 kg Aberglyn tetraploid perennial ryegrass, 5 kg premium perennial ryegrass, 7 kg Portrush perennial ryegrass, 6 kg Ideal tetraploid perennial ryegrass, 5 kg Promesse timothy, 0.5 Aberherald white clover, 0.5 Crusader white clover and 0.25 Aberpearl white clover.	An excellent dual purpose ley. A combination of intermediate and late perennial ryegrass, the timothy and clover improve palatability. Ley is robust and can be cut for silage or intensive grazing with cattle and sheep.
Traditional grazing and hay (on light soils)	7 kg Prairial cocksfoot, 12 kg Aberglyn tetraploid perennial ryegrass, 7 kg Rossa meadow fescue, 3.1 kg Erecta timothy, 1.2 kg Aberherald white clover and 0.6 kg Abercrest wild white clover.	A good and durable traditional ley. The inclusion of the cocksfoot, timothy and clover ensures a good palatable grass ley. The deep root system of the cocksfoot allows moisture to be drawn up from great depths and ensures long term durability on light soils. This ley will withstand continuous grazing, indeed required to prevent cocksfoot developing clumpy tussocks, and is suitable for a good hay crop.
Deep rooted forage ley for dry land	8 kg Prairia cocksfoot, 8 kg Rossa meadow fescue, 5 kg Erecta timothy, 1.2 kg Aberherald white clover, 0.6 kg Abercrest wild white clover, 3.1 kg of commercial Sainfoin, 1.2 kg Vela alfalfa, 1.2 kg Chicory forage herb, 1.7 kg Burnet forage herb, 0.5 kg Ribgrass forage herb and 0.25 kg Yarrow forage herb.	On very light land and land that dries out during summer conventional ryegrass leys will burn off (dry up) reducing the area of summer grazing. On such lands and areas where low input farming is desirable this seed mix offers a flexible summer grazing regime. The high number of species in this mix ensures durability and palatability, the inclusion of the Sainfoin and clover helps to maintain soil fertility.

Notes: Seed rates a rounded and indicative. All seed is certified.
Adapted from www.thegrassseedstore.co.uk and www.cotswoldseeds.com, accessed November 2009.

Table 3.2 Grassland year

Time of year	Action	Comments
Winter	House stock, especially cattle to reduce poaching. Apply lime if required. Check land drains, repair hedges and clear ditches. Apply FYM in late winter (early February)	No fertiliser should be applied. Keep heavy machinery and stock off water-soaked grassland.
Early spring	Harrow pasture to remove dead material (thatch) and to pull up shallow rooted weeds. Roll pasture to level molehills and consolidate the grass. Apply FYM where required.	Decide on which field to be closed for haymaking and fertilise accordingly.
Mid-late spring	Turn stock out to graze. Restrict access to whole field by strip grazing. Control access with electric fence. Top rejected material and weedy paddocks.	Watch for signs of scouring and hypo-magnesia.
Early summer	Cut hay as grass sward starts to flower. Continue to control grazing. Where grassland is showing signs of extensive weed growth, top off before seeds set.	–
Mid summer	Take a second cut of hay if required. Top off weedy paddocks if necessary.	–
Autumn	Continue to control grazing while pasture is dry. Once heavy rains have soaked pasture, move stock to winter quarters.	–

grazing, then these marginal lands can be utilised, nutrient and pesticide inputs reduced, high management costs reduced, biodiversity increased and a wider range of high quality products can be put on the table. Later in the chapter we discuss a number of cases where grazing is used as a habitat management strategy.

Owing to the importance of managed grazing to the whole ethos of conservation on farmland, in this section we look in some depth at a number of breeds of cattle, sheep and ponies that are currently being used in the UK in conservation grazing programmes. It should be noted that the principles are just as pertinent to grasslands globally.

Choosing the right breed for a given grazing system is not easy and breed selection must consider the commercial value of stock as well as the appropriateness for any planned grassland management. Generally the modern livestock farm is based around the black and white Holstein for dairy, Semimetal/Limousine for beef, with local variations including the South Devons and Ruby Red. Sheep production systems are based on the Mule or Cheviot for lamb production and in Kent, UK, especially the Romney Marsh; sheep production is based on the Kent, which is a large breed. All these modern commercial breeds require expensively maintained high quality

Table 3.3 Inorganic fertilisers

Production system	Nutrient	P or K index			
		0	1	2	3
		kg/ha/yr			
Grazing by dairy cows	Phosphate (P_2O_5)	60	40	20	0
	Potash (K_2O)	60	30	0	0
Silage (first cut 23 tonnes/ha fresh weight)	Phosphate (P_2O_5)	90	65	40 M	20
	Potash (K_2O) previous autumn	60	30	0	0
	Spring	80	80	80 M (2−) 60(2+)	30
Hay	Phosphate (P_2O_5)	80	55	30 M	0
	Potash (K_2O)	140	115	90 M(2−) 65(2+)	20
		Soil nitrogen supply			
		Low	Moderate		High
		kg/ha/yr			
Grazing by dairy cows (28 day grazing, high stocking rates)	Nitrogen	380	340		300
Silage 69 to 70D value					
First cut		150	120		120
Second cut		110	100		100
Third cut		80	80		60
Hay		90	70		60

pasture, which can incorporate a three-cut silage production system for winter feed. This intensive approach requires a significant input of inorganic fertilisers (Table 3.3) and the rotation of short-term leys (less than five years) all of which is in conflict with the current philosophy of grassland conservation and may be inappropriate for many semi-natural and up-land systems which have less productive pasture.

The extent of semi-natural grassland in the UK is unknown. Significant tracts of semi-natural grassland occur in National Parks, which lie within the upland communities of the UK, including Dartmoor. Other grassland communities (such as calcareous grasslands) occur within the boundaries of commercial farms and thus all of these grasslands have a financial implication for the land manager. Many upland farms are small and have limited scope for cropping (Case Study 3.3 below), and thus the mainstream income will come from their livestock operation. On the larger lowland farms livestock may not be the mainstay of farm income but any successful business will need to ensure all its commercial activities at least break even, and breed selection, therefore, will need to focus on those breeds that can perform well on less intensive pasture.

The producer needs to control production costs (fertiliser, feed and vets' bills) and maintain a good income from their stock; this can be achieved by considering some of the traditional breeds of sheep and cattle that have lost favour with the modern commercial farm. Traditional breeds tend to be smaller, perform well in less intensive pasture and are generally hardy and thrifty compared to modern commercial stock. They can also reduce costs from illness as many breeds have a natural immunity to tick borne disease such as red water and tick fever. Furthermore, the producer can add value to their product by selecting rare breeds that are in demand on the niche meat market, their wool for spinners and weavers and their hides for the leather trade. Value can also be added by selecting stock that has a demand because of their pedigree, which in the modern rural scene can find a role in conservation grazing schemes.

Conservation in rural Britain is achieved in many ways from on-farm conservation through recent European initiatives arising from CAP reform such as; Countryside Stewardships Scheme (CSS) agreements, Sheep and Wildlife Enhancement Scheme (SWES) operated by Natural England and more recently (since 2005) a rolling programme of agri-environment payments, which will replace CSS and from 2010 the Uplands Entry Level Scheme (UELS) will be introduced. Habitat and species conservation is also achieved through many non-governmental organisations (NGOs) such as the Royal Society for the Protection of Birds (RSPB) and Country Wildlife groups and of course the National Trust. All these groups have significant areas of pasture that are being actively managed through the use of suitable grazing animals, which selectively control the species composition of the grass sward.

3.3.1   Grazing animal profiles

The three main domesticated grazing animals are cattle, sheep and ponies each with a distinctive pattern of grazing and feeding preferences. Cattle have long and rough tongues that they wrap around tall vegetation and through a combination of biting and pulling feed on long grasses and thus cattle have difficulty in grazing short swards. This feeding behaviour produces a relatively high and uniform sward. Sheep, on the other hand, bite and nibble and produce a tightly grazed low sward, often-ideal characteristics of calcareous grasslands where low grasses enable a high density of small forbs. However, prolonged sheep grazing can seriously degrade heavily stocked sites. Ponies forage a range of grassland species and unlike sheep and cattle, ponies will graze at varying heights producing structural diversity in the sward, a key element of successful grassland management for diversity in faunal wildlife.

The breeds presented here have been selected for their hardiness, thriftiness and for their potential added value.

3.3.2   Cattle

Cattle breeds have been loosely grouped into three categories according to their position in the production system:

- dairy;
- beef;
- dual purpose.

Aberdeen angus

A predominantly black, polled, beef breed which is hardy and easy to handle. The breed is able to cope with slopes and can graze a range of grassland from chalk downland to lowland marsh and to upland acid pasture. Aberdeen Angus were developed in north east Scotland in the early nineteenth century by Hugh Watson of Keillor by crossing two local black breeds, the Hummlies and Doddies. The breed has a reputation for quality beef and has been consistently improved since the nineteenth century and is a favourite breed on low input systems. The breed is used by Surrey Wildlife Trust, the National Trust in Wiltshire and the RSPB in Essex. The breed has good commercial characteristics and the calves market well.

Belted galloway

An attractive and very hardy breed (Figure 3.2) that can adapt to a wide range of habitats and environmental conditions. Particularly useful in wet climates and in

Figure 3.2   Belted galloways. (Photo: Stephen Burchett)

year-round grazing situations. This is a gentle and placid animal, which makes it a very useful animal for grazing sites with public access.

Summary of characteristics

This breed is not ideally suited to being housed during winter months and is unlikely to grow a winter coat in this situation.

- **Supplementary feed:** fares well on coarse grasses even during winter and may only need minimal amounts of hay or concentrate.
- **Handling and breeding:** generally docile and easy to handle, cows are good mothers producing a strong supply of milk for the calf.
- **Sure-footed:** is particularly adept at grazing on steep slopes without causing poaching or erosion damage. Flatter areas are necessary for resting and ruminating.
- **General health:** very good and not particularly susceptible to insects such as ticks and fly strike and thus can escape tick borne disease. Cows will usually live to around 12 years of age with an average size of 500–600 kg.
- **Grazing characteristics:** adaptable to grazing and browsing a great range of species and habitats, not particularly selective and appears to take a wide range of grasses, shrubs and coarse herbs and shrub species (Table 3.4) (GAP, 2001).

Highland Kyloe

The quintessential icon of the Highland rural community appearing on many tourist souvenirs is the long horned dishevelled coated cattle (Figure 3.3) known as the Highland Cattle (Kyloe). This is an extremely hardy and intelligent breed that performs well with quiet handling and can cope quite successfully with the harsh conditions of its extensive Highland range. The breed does not fair particularly well when housed over winter.

The Highland breed has a dual layer coat consisting of a waterproof, long, coarse haired outer coat and a short down inner coat for warmth; the outer coat is shed during the summer months. The benefits of this dual layered coat are that this breed can withstand prolonged spells of harsh driving rain and intense winter chill making it an ideal breed for upland grazing. The extensive and isolated home range of this breed results in herds becoming semi-feral with pregnant cows calving unaided. The cow will detach herself from the herd and select a sheltered and secluded location, often by streams in deep grass and ravines, and once the calf is born the cow will leave the calf in tall grass and rejoin the herd for grazing. Once satiated the mother follows a scent trail back to the calf.

The ranging behaviour of this breed is reminiscent of true wild herbivores. However, the breed can be domesticated and handled by gentle coaxing using supplementary

Table 3.4 Conservation grazing using belted galloway

Site	Comments
Rodorough common SSSI Gloucestershire National Trust. Limestone grassland common land with steep slopes – 115 ha	Nine+ belties were introduced in 1999 to help manage botanically rich sites throughout the South Cotswolds. Winter grazing on steep slopes using electric-fenced paddocks. Cattle have successfully removed thatch, tor-grass, upright brome as well as suppressing/eating cotoneaster, scots pine, holme oak, silver birch, blackthorn, hawthorn, ash and whitebeam. In summer, a variety of other wildlife sites grazed mostly limestone grassland but also calcareous marsh.
Boxmoor Trust Hertfordshire. A wide range of habitats including chalk grassland arable reversion and wet meadow	Belties introduced in 1995 and used at different times of year. Stock start to graze the fescues and timothy leaving the bents and cocksfoot until last. The stock will browse a wide range of chalk grassland shrubs including hawthorn, cherry, young ash, beech, blackthorn and spindle. Wayfaring tree tends to be left until last. They will browse the juniper if left on site when all other palatable vegetation has been taken but they are taken off before this stage as the trust does not want the juniper to be browsed. A good impact on bramble, largely through trampling.
Trendlebere Down Dartmoor ESA. Upland moor with lowland heath interface. Bracken and gorse present. Common land	Ten belties of mixed ages introduced in 1999, grazing March to September to address decline in grazing on common. Some sheep and ponies also on site. Belties moved to similar habitat on the farm during winter, in preference to South Downs which cause poaching damage. The area is recovering from a fire in 1997 and the cattle have been found to graze well and unselectively. On leaving the moor homed in on roadside nettles and docks, species not found on the moor itself. Good trampling of bracken and bramble in lying-up areas and pathways.

Figure 3.3   Highland Kyloe. (Photo: Stephen Burchett)

feed such as hay. The semi-feral nature of this breed means it will challenge boundary structures and can quite easily push over poorly erected post and wire fences, overcome water filled ditches and cattle grids. The most successful boundary is a low wall.

The horned nature of this breed means that handling should be done with great respect, indeed the breed will generally ignore most people, but cows with calves at foot will be protective and will successfully challenge any dog that threatens the calf. Bulls are seldom problematic if kept with a herd.

Generally this is a healthy breed requiring limited medical intervention compared to modern commercial breeds, some problems can occur with mange and fly strike due to the long coat. The breed performs well on coarse vegetation but requires a mixed diet with lots of bulky herbage for good performance. Old cows can become arthritic.

Grazing animals will take grasses and herbs, including coarse grass and can make a significant impact on lank purple moor grass (*Molina caerulea*). In wetter areas the breed will selectively graze reeds in spring and early summer and likes reed sweet grass (*Glyceria maxima*) but dislikes saw sedge. During late summer cattle will take nettles (*Urtica spp.*). The browsing activity of this breed results in good scrub control with cattle actively taking willow (*Salix spp.*), but it does not selectively graze heather and other ericaceous species unless fodder is in short supply. Herds adopt a home range, which can be several square kilometres, and will graze in a cycle returning to the origin after several days of ranging. This dispersal of animals reduces the negative impacts of grazing. This breed is used in conservation grazing and is a favoured breed for fen and wet meadow (Table 3.5).

Table 3.5 Conservation grazing using highland Kyloe

Site	Comments
Yare broads and marshes Norfolk RSPB Wet tidal eutrophic fen scrub	A small, non-commercial herd was introduced in 1995 to graze this difficult site. Animals roam freely and willingly wade into treacherous wet pasture. Grazing was introduced to control willow and reeds and reed sweet grass. Minimal poaching has been observed but handling can be problematic and because of standing water liverfluke need to be controlled with a worm drench.
Bure broads and marsh Norfolk Natural England A species-rich fen on deep peat with scrub	Introduced in 2000 to control scrub encroachment by year-round grazing. Higher drier pasture is available for winter housing. Contained by electric tape fencing, dykes and wet ditches easily waded across.

Markets for the Kyloe lie in the sale of good pedigree stock, especially good show standard females. The meat market is niche but the meat contains low levels of saturated fats and due to its extensive range and slow growth rate the final meat product is of high quality.

The breed is also used in its native range for conservation grazing, for example the Glenborrodale Estate on the Ardnamurchan peninsular in the Highlands of Scotland has a commercial herd of Kyloe and animals can be readily observed across this rugged peninsular grazing the coarse grasses and helping to maintain the heather moors as well as providing an important sustainable venture for local communities (Figure 3.4).

Figure 3.4 Small highland Kyloe herd. (Photo: Sarah Burchett)

The White Park

The White Park is a horned breed (Figure 3.5) which is white with dark coloured points, these points include ears, nose, rims of eyes, teats and feet but excludes the tail switch. The upper portion of the tongue should be black while underneath is most often pink. The intensity of these markings varies from herd to herd. The White Park is now considered a beef breed and selected for those traits but historically they were considered dual purpose since some herds have historically been used for milk production. It is reported to be well adapted to non-intensive production systems, (www.Whiteparkcattle.org. – accessed 14 November 2009).

The White Park is the oldest registered breed in the UK dating back some 2000 years. During their long history White Park cattle have been used for many purposes:

- They were used as a special form of currency
- Probably were important as sacrificial animals and as beasts of the chase
- They were used as draft animals in the nineteenth century, there is a record of the last plough oxen in the Dynevor herd being slaughtered in 1871 at 14 years of age, when he stood '6 ft at the withers and weighed 23 cwts (approximately 1168 kg) and his horns measured 5 ft (1.5 m) from tip to tip'
- In the early twentieth century some herds were milked.

Until recent years the majority of breeders paid little attention to the commercial qualities of the breed, but there is now an increasing awareness of its potential value provided that positive development policies are followed.

Figure 3.5   White park cattle. (Photo: Stephen Burchett)

Ease of calving

The White Park experiences few problems at calving either in pure breeding or when used as a crossing sire. A project carried out by the Rare Breeds Survival Trust (RBST) demonstrated that the White Park compares very favourably with other breeds for ease of calving in pure breeding. Results from several commercial herds show that White Park bulls cause very few problems if any when used as a crossing sire, and the crossbred calves are notably active at birth. Calves are left in nursery groups.

Thriftiness

The breed exists in a wide variety of conditions from lowland meadows with housing in winter, to out wintering on Pennine pastures. It has the ability to grow well on much poorer feeds than modern continental breeds keeping feed costs down. It is reported to be well adapted to non-intensive production systems (Mr Lean, Personal communication).

Size

The White Park is a large breed, with cows weighing on average around 600 kg (12–12.5 cwt). It is also long in the hindquarters (hip to pin), and this is of value not only for ease of calving, but also for beef production.

Breed genetics

The White Park is a genetically distinct breed and is free from deleterious genes found in many other British breeds. The breeds that appear to be most closely related to the White Park are the Highland and Galloway cattle of Scotland. The genome of the White Park is so diverse that it has evolved several resistant traits that make it ideal for low cost production systems. It is resistant to a lot of parasites found in pasture and stream-fed water supplies.

Present status of the herd

The current status of the White Park is critical with a breeding population of approximately 600 breeding cows worldwide (www.Whiteparkcattle.org – accessed 14 November 2009). The White Park cattle in the US have been DNA typed for purity and to determine the best breeding plan possible to save its genetic base. An ongoing breeding programme has been put into place to help ensure the breed's survival. The breed reached its most endangered position following the Second World War. During the war the White Park was considered by the government to be sufficiently important as part of the British heritage for a small unit to be shipped to the USA for safekeeping. During the 1960s only four domesticated and recorded herds remained, namely the Dynevor, Cadzow, Woburn and Whipsnade herds. Since 2005 a herd of 200 breeding cows has been established by Sir Benjamin Slade, the Maunsel Herd, which is now the largest herd of White Park Cattle in the UK (www.Whiteparkcattle.org – accessed 14 November 2009).

The current series of Herd Books were collated from 1972 onwards, and full records on all animals have been maintained since that time. However, detailed records had been maintained in some herds for a much longer period. The breed is officially recognised in the EEC, and animals have been exported to North America, Australia, Denmark and France. The White Park Cattle Society is responsible for maintaining the herd book and protecting and overseeing the development of the breed. Pedigree White Park Cattle must be registered in the herd book, and meat sold as 'White Park Beef' must be from offspring of pedigree parents.

Beef shorthorn

This breed is steeped in the cultural history of UK livestock, originating 200 years ago from a cross between the Durham and Teeswater cattle and was essentially a dual purpose breed, providing a role as a dairy cow or beef stock. Cattle can be either horned or polled and produce a good finished animal between 24 and 26 months old weighing 300 kg at market. Cows continue to produce good calves for many years, typically producing 10–12 calves in a lifetime compared to the three or four calves produced by many modern commercial breeds. This fertility keeps replacement costs down. The colouration of this breed is quite distinctive, ranging from red, red-and-white to white-and-roan. This is a calm and placid breed that has a commercial market and is used in several conservation sites such as Hod Hill in Dorset (National Trust) and by the RSPB at Lodmoor and Radpole Lake in Dorset. A ready market is available through the RBST.

The longhorn

The longhorn is another breed with ancient origins and quintessentially the beast of burden and draft oxen of ancient Britain. Interestingly they are still used today to draw felled timber in sustainable woodland projects such as the Maine community woodland project. The modern longhorn has its origins steeped in the 1700s and is a result of crossing two improved heifers and an improved bull from Westmorland, which was a descendent of a line of inbreed cattle aiming at fat production over bone and milk (Hall and Clutton-Brock, 1989). The result of this cross is the breed we see today, which is a hardy and adaptable breed with good grazing characteristics and a finished weight from 250 to 381 kg. This breed is thrifty and maintains condition well on rough pasture, ranging in colour from red, brown, grey, brindle and varicoloured. Both sexes are horned with individual horns on bulls reaching 60 cm in length; this results in an intimidating appearance (Figure 3.6), but the breed is docile and graceful, indeed a magnificent animal. The breed lost favour by the mid 1800s as the improved characteristics of the shorthorn was readily adopted by the industry. The National Trust at Cheddar Cliffs, Somerset, UK, uses the longhorn for summer grazing on around 19 ha of conservation grassland.

Figure 3.6 Longhorn cow. (Photo: Stephen Burchett)

3.3.3 Sheep

Sheep are generally classified into two main groups according to their regional production. The lowland sheep are, in the main, large sheep requiring high quality pasture; the mule can be classified in this group but can also thrive in the uplands. The mule is a hybrid animal resulting from crossing a hill breed, like the Scottish black face or a Welsh mountain ewe with a Bluefaced Leicester, a short-wool hill breed. The offspring are hardy and thrifty with high fertility and good mothering instincts, features that are selected for by shepherds.

Lowland sheep were developed for wool and meat production and typical, traditional breeds are Dorset and Hampshire Downs and on the Romney Marsh, in Kent, a large long-wool sheep (80 kg ewes) predominates, known as the Romney or Kent sheep. This is an ancient breed dating back to Romano-Britain (Hall and Clutton-Brock, 1989), with the breed being improved in the eighteenth century by Richard Goord of Sittingbourne (Hall and Clutton-Brock, 1989). The Romney sheep has significantly influenced the global sheep industry, in particular New Zealand, Australia and South America where 18 000 rams and 9000 ewes were exported between 1900 and 1955 (Hall and Clutton-Brock, 1989).

The next group is the **hill sheep** which have been selected for their hardiness and ability to thrive on poor quality grasslands and hill pasture; these tend to be small and undesirable for lowland production systems and thus have been marginalised by contemporary agricultural systems. However, they can be extremely useful in conservation grazing systems. Many of these sheep are registered with the RBST.

Jacob

This is an ancient breed. Able to remain outside all winter in the south, but in periods of heavy rain they should be housed indoors as the fleece is not actually waterproof. Except in poor weather the breed generally maintains condition well on poor quality forage. The breed is tolerant of most grasses but does not like dank/dead swards. There is a niche market (speciality restaurants) for the dark lean meat but the lambs are slow to finish, the multicoloured fleece can be sold direct to spinners and weavers. This breed suffers from fly strike to the head and the open fleece. Jacobs have been used at Arlington Court in Devon (National Trust) and at Pentire Head in Cornwall (National Trust) and on several other conservation projects in the UK (Figure 3.7).

Portland sheep

These are an RBST endangered species with around 2000 sheep in 100 small flocks throughout the UK. Portland sheep are renowned for their thriftiness and do very well on marginal land – two Portland sheep can be sustained on the same area of land that a commercial ewe would occupy. The ewes are small averaging 38–40 kg. They generally produce one lamb that fattens in around 12 months. The lambs are known for their quality mutton and their meat was once considered a delicacy by the gentry. The ewes can be crossed with a terminal sire such as a Southdown to produce a commercial lamb. Both sexes carry horns that can develop a distinctive black line. The lambs are generally tan coloured indicating that the breed has ancient origins. The Portland sheep offers many added benefits such as a highly desirable fleece, which

Figure 3.7   Jacobs ewe with lamb at foot. (Photo: Stephen Burchett)

is in demand by hand spinners, and the horns are in demand from stick makers. The hide is ideal for leatherwork.

Shetland

This is a hardy prolific and thrifty breed that will generally produce two lambs that will weigh from 20 to 38 kg at about six months old. Their meat is lean and of high quality. This breed thrives on a low intake of feed and can succeed on poor grassland. Their fleece is multicoloured and thus the wool commands a premium and can be sold direct to spinners. They are used by the Norfolk Wildlife Trust as a 'flying flock' for summer grazing on dry and wet heath and fenland.

Wiltshire horn

This is a minority breed with a very short fleece with an established commercial market. The lack of fleece reduces the risk of fly strike. Tolerant of hot and cold climates, this breed was developed on the exposed hills of Wiltshire. It produces lean meat of good flavour, the lambs are quick growing, the rams are popular as terminal sire and ewes can be crossed with terminal sire for lamb production. They have been used by Dorset Wildlife Trust on chalk grassland and by English Nature on Old Winchester Hill.

Hampshire down

Established 150 years ago by crossing the Southdown with traditional mutton producing downland sheep, of lowland Britain, to produce the improved Hampshire Down breed. This breed was developed to produce a sheep that could be adapted to changing land use of the Hampshire Downs in the 1800s. During this time much of the traditional downland sheep walks were being ploughed and sown to cereal crops, changing the structure and nature of lowland farming from a traditional sheep industry to an early model of mixed farming, incorporating forage and fodder crops and the use of artificial manures. This new approach to downland farming initiated the development of the fresh lamb market over the traditional mutton trade. The Hampshire Down breed is eminently suited to this role. Hampshire Down sheep are a dark faced hornless breed where the fleece extends over the head covering the forehead and cheeks. This is a sure footed breed with wide set stout legs producing robust lambs that perform well on mixed grasses, finishing early with a good meat quality and high grading out percentage.

3.3.4 Ponies

Two breeds are highlighted here, the Dartmoor and Konick ponies, but there are many other hardy hill ponies such as the Dales and Exmoor pony. Ponies are ideal animals for many conservation projects, for example Dawlish Warren use Exmoor ponies as part of an integrated scrub management strategy.

Dartmoor

This is a small and hardy breed originating from the moorland and mires of Dartmoor and survives well on marginal grassland in harsh conditions. However, the bloodline has been diluted over the years with crossbreeding programmes reducing the breed's ability to survive in the exposed conditions of Dartmoor. Modern breeders are attempting to reduce the impact of historical crossbreeding programs by selecting individuals who demonstrate the features that resemble the original breed line such as colour which should be bay, brown or black (Figure 3.8).

Physical attributes and husbandry

A medium-sized animal of some 350–450 kg with well muscled quarters and a heavy coat that enable the breed to cope with exposed locations. The Dartmoor can be domesticated and is a curious breed, which with handling will become an ideal pony for the in-keep. Animals kept in small herds tend to be friendly towards the general public but individuals can be flighty and problematic. Demands on husbandry are minimal although worming is recommended and if domesticated and removed from a semi-natural environment, then husbandry requirements increase with the need to trim feet and provide shelter such as shade from intense summer heat. The major

Figure 3.8   Dartmoor pony. (Photo: Stephen Burchett)

insect problem is fly and tick issues and therefore regular inspection is required to ensure conditions such as these parasites do not cause any welfare problems.

Grazing characteristics

Grasses dominate diet with favourites such as cocksfoot, fescues and purple moor-grass with a general dislike of bristle bent grass and false oatgrass. Smaller sedges and soft rush are readily taken and the breed will browse brambles and if short of grass may switch to feeding almost entirely on gorse with browsing on heather tops if pushed. The Dartmoor can result in some suppression of willow, birch and blackthorn regrowth.

The Dartmoor does not normally select coarse herbs and flowering, but may occasionally nibble leaves of thrift and primrose. It is an adventurous animal and pushes into, and opens out areas of dense vegetation, including bramble and bracken.

The Dartmoor is used in a number of conservation schemes (Table 3.6) which include the loan of ponies to conservation groups by the Dartmoor Pony Society (DPS).

Konick polski

An extremely hardy breed that is capable of out-wintering in temperatures as low as $-40\,°C$. An intelligent animal, the Konick (Figure 3.9) makes seasonal dietary choices, actively ranging to seek food items and shelter. In the UK the Konick has been mainly

Table 3.6 Conservation grazing using Dartmoor pony

Site	Comments
Castle Drogo National Trust, Devon Steep sided woodland gorge with heath and scrub	Three to seven ponies graze the gorge from March to October opening out pathways and reducing the impact of scrub on the violet population, thus maintaining population of heath fritillary. There is a definitive switch of diet from grasses to gorse in late summer. The National Trust at Castle Drogo have noticed that the Hornet Robberfly uses th dung of the ponies.
Hembury woods National Trust, Devon Heathland, scrub and grassland	Up to 12 ponies used since 1992. Ponies graze most grasses leaving Bristle Bent until last. They trample bracken and take bramble, they will eat gorse and heather is pushed. The purebred ponies maintain well at times compared to crossbred ponies. Handling is difficult and worming is carried out in late summer.
Pencarrow head, Lansallos Cliff and Valley. National Trust, Cornwall Neutral coastal grassland scrub	Thirty ponies are used across five different sites from November to April since 1994. Ponies have created a mosaic landscape, opened up pathways and reduced the dominance of course grasses. Winter grazing is now resulting in poaching of soils and a switch to summer grazing is planned.

Figure 3.9   Konick Polski ponies. (Photo: Stephen Burchett)

used in wetland grazing schemes but the foraging behaviour of this animal makes it suitable for many other open semi-natural grassland grazing projects.

Grazing characteristics

Well adapted to making appropriate food choices in extensive areas readily taking a range of species, including elm, willow, oak, hawthorn, brambles and wild privet on coastal dunes, alder and birch are generally ignored, it may actively seek wood if other browse material not available. Some individuals will take mature trees in preference to young scrub, which is an important part of diet in year-round grazing systems. This pony is effective in controlling the invasion of scrub into open habitat.

During spring and summer the Konick prefers grasses but by September, takes more varied diet if available including sedges, rushes, seed heads of thistles and reeds. It will also dig up and eat roots/rhizomes in winter, including those of the stinging nettle and common reed. This breed will explore its territory and adapt its diet accordingly to include new food items not previously encountered. Such pony grazing at Stelling Minnis in Kent, UK has resulted in a mixed structured sward, which has a diverse range of floral species including many rare orchids.

The above is just a small selection indicating how grazing animals have contributed to the creation of semi-natural landscapes in the British Isles. It is the intensification of agriculture and the pressure to produce more food on the same area of land that has resulted in the development of modern commercial breeds. These breeds require high quality nutritious grasses, based on the perennial ryegrass sward, that have brought about the decline in species-rich pastures. Now the threat to these species-rich pastures comes from the demise of agriculture and the knowledge required to successfully keep large animals. The future clearly lies in the return to traditional mixed farming and the development of local markets for quality produce sold at the correct price, there is no such thing as cheap food. In Britain, imports of beef from Brazil are at the expense of the tropical rainforest and result in increased deposits of atmospheric carbon and loss of British farming in marginal areas such as the uplands. Using breeds that thrive on minimal or no inputs clearly is beneficial to the farmed environment but is also economically viable for the landowner.

3.4   Making more of native grasslands

3.4.1   Grasslands of the United States

The American continental landscape is dominated by diverse grassland habitats. Here we highlight some of these habitats with respect to farming and to the conservation issues and initiatives associated with them.

Sagebrush

Sagebrush (*Artemisia tridentate*) forms a unique semi-arid 'steppe' type habitat that is found across many parts of western United States and Canada and supports a number of threatened species such as the sage grouse (*Centrocerus urophasianus*) (Figure 3.10). In Montana the Natural Resources Conservation Service (NRCS) has targeted the sage grouse as a key species in its conservation management initiative. Prescribed grazing with cattle is one of a number of tools used to manage the sagebrush habitat that the sage grouse relies on for food and its ground nesting sites. Sagebrush is unpalatable to most animals though the grouse thrives on it; the cattle graze the grasses and forbs around the sagebrush reducing potential competitors for ground space, particularly pertinent in light of increasing air moisture possibly indicative of global warming effects. Much of the sagebrush steppe in Montana is privately owned, therefore grazing cattle here has the added advantage of supplying revenue to the landowners without the need to plough the land for crop production. Other species reliant on sagebrush steppe habitat are the brewers sparrow (*Spizella breweri*), the sage sparrow (*Amphispiza belli*), the sage thrasher (*Oreoscoptes montanus*), the pygmy rabbit (*Brachylagus idahoensis*) and the sagebrush vole (*Lemmiscus curtatus*) (www.mt.nrcs.usda.gov – accessed July 2009).

Prairies

We were unable to visit any prairies during our travels, however, we regard this biome as one of the most important for wildlife conservation within farming in the US and indeed in the world. Rescued from the brink by a formidable collection of American, Canadian and international stakeholders, prairies represent the epitome of what the

Figure 3.10   Sagegrouse. (Photo: Sage Grouse © Bruce Waage – NRCS)

global conservation effort is all about. For this reason it has been included in the suite of case studies on grasslands.

Case Study 3.1: Prairies

Prairies are temperate lowland grasslands occurring in North America and Canada. These habitats have been largely destroyed by ploughing for cereal production, and yet they still occupy vast areas of land. Once these grasslands stretched from the Great Plains all the way to the mid-west. Only 1–2% of prairies now exist, small pockets as 'virgin' prairie and the rest as restored prairie, but this still constitutes approximately 3.6×10^8 ha (1.4×10^6 miles^2).

The Formation of Prairies

There seems to be some disparity in the varied texts as to how the prairies were formed. Some say that this is a semi-natural habitat that developed after European settlement and mass deforestation (McCracken Peck, 1990). Others state that it is a natural habitat formed after the last glaciation, where, in the first instance, the soils were too poor to support tree growth. In reality there are undoubtedly elements of fact in both scenarios, given that it is such a vast area. With such variation occurring in the habitat existing under the heading of 'prairie' there is unlikely to be just a single causal influence. Indeed, in the more easterly regions, where rainfall is higher, deforestation for agriculture would have had a major influence, in the drier more westerly regions prairies are influenced by the Rockies which act as a weather barrier, these grasslands fall into the 'rain-shadow' of the mountains.

What is certain is that low rainfall, grazing and frequent fires have prevented tree sapling growth.

The Prairie as a Habitat

There are different types of prairie; rainfall is lower further to the west, usually around 30–35 cm, so that different grassland species (grasses and herbaceous plants or forbs) are to be found which best exploit such environmental conditions. The west mainly supports short-grass prairies, central parts have mixed-grass prairies and the eastern prairies, where rainfall is normally around 50–55 cm, are mainly dominated by tall-grass species that can grow up to around 3 m. Southern prairies also support the taller species influenced by warm moist air blowing up from the Gulf of Mexico. Map 3.1 shows the area where prairies exist, though it should be remembered that prairie grassland now occurs in fragmented pockets within these regions.

Prairies are maintained largely by grazing and by fire, very few trees are found in this habitat as every few years the dry grasses catch alight in the hot summers and are blown far and wide by the prevailing winds. These fires prevent growth of tree saplings and burn grass thatch, helping to fertilise the soils and giving space for prairie flowering plants, such as phlox, viola, indigo (*Baptisia spp.*) and milkweed (*Asclepias spp.*), to proliferate the following spring. Thus prairie flowers tend to be dominated by deep-rooted species with perrenating organs such as bulbs and rhizomes. Also, the apical meristems are often found slightly below the ground surface to avoid damage from the burn and from grazing. (www.bluepanetbiome.org – accessed October 2009) These deep rooted plants also have a tangled network of secondary roots forming dense mats which help to prevent soil erosion across the plains, and which add to the cycle of organic matter production.

As prairies are found in temperate regions where the summers are hot or warm and dry and the winters are cold or cool and wet, they support a vast diversity of animal species including mammals, such as the plains pocket gopher (*Geomys bursarius*) and the coyote (*Canis latrans*); reptiles including a number of snake species such as the western hognose (*Heterodon nasicaus*), the prairie king snake (*Lampropeltis calligaster*) and the prairie rattle snake (*Crotalus viridis*), as well as a vast array of invertebrates. Many birds occur to include both resident species, such as the greater prairie chicken (*Tympanicus cupidus*), which is highlighted in Species Box 3.1, and migratory birds, such as the bobolink (*Dolichonyx orzivorus*) and the grasshopper sparrow (*Ammodrammus savannarum*) (www.americanprairie.org – accessed October 2009). Lack of water would be a limiting factor, but the prairies are dotted with wetland areas formed when the glaciers receded around 10 000 years ago leaving depressions that fill with water when the rains come (www.globalchange.gov – accessed October 2009).

Pollution is a significant factor. When the prairies covered vast tracts of land across North America large mammals such as bison followed ancient migratory routes. Now these prairies are fragmented and criss-crossed by highways and urbanised areas. The herds of bison, elk and pronghorn (*Antilocapra americana*), – see Species Box 3.2 – which still exist and are a recognised conservation 'at risk' species, but now they are contained within boundaries. Their grazing and their deposition of waste products are therefore also contained within these boundaries changing the dynamics of nutrient cycling both in the areas where they live and in the areas that they are no longer passing through.

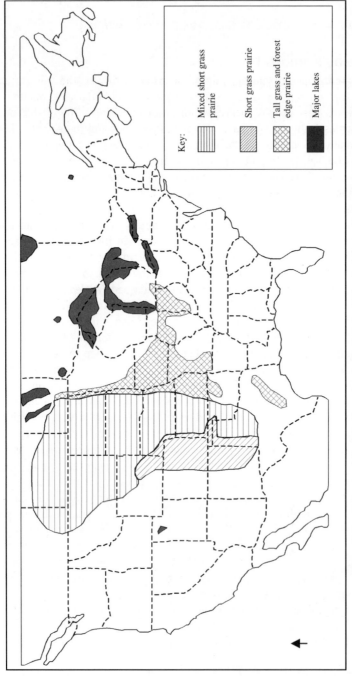

Map 3.1 Extent of Prairie Grasslands in U.S.A. (Adapted from McCracken Peck, 1990)

Species Box 3.1: Prairie Chicken (*Tympanicus cupidus*)

Profile:

Photo: Prairie Chicken © Don Brockmeier

The prairie chicken lives exclusively in grassland in North America and Canada. Once, prairies covered vast tracts of land throughout this area, and with their decline also came the decline of species dependant on them such as the prairie chickens. They require rich forage, preferably including legumes such as clovers, and good winter stubbles. Tall grasses of at least 20–25 cm by mid-April are needed for them to breed. The habitat landscape needs to be around 75% suitable grassland to support a good viable population. It feeds on seeds, fruit insects and sometimes green leaves.

Conservation Status:

Tympanicus cupidus (T.c.) Three Sub-Species

1. *T.c. pinnatus* (greater prairie chicken or pinnated grouse);
   (a) IUCN 1996 Low risk.
2. *T.c. attwateri* (Attewater's prairie chicken) IUCN red list, vulnerable'
   (a) CITES 1998 Appendix 1.
3. *T.c. cupido* – now extinct.

Current Status:

Now numbering less than half a million individuals and still declining, the prairie chicken has very specific needs to recover and to survive. A number of organisations, including the Missouri Department of Conservation have laid out procedures for farmers to ensure the continuity of the species. The few remaining wild prairies and cultivated native warm-season grasses will support prairie chickens. Grazing rotation with fallow seasons, plus use of native grasses with regular burns, and with legumes in the seed mix, ensures good foraging for both the prairie chickens and for the grazing stock. Regular burns of native grasses removes woody growth not favoured by the chickens, and it improves harvest quality for the farmer. Cutting can be done in July, but if left to autumn, the regrowth will not reach the critical height for the prairie chicken breeding season.

References

www.allaboutbirds.org (In association with Cornell University)
www.nature.org

Species Box 3.2: Pronghorn (*Antilocapra americana*)

Profile:

Small ungulate ruminant, approx. 1 m shoulder height, endemic to the American continent and living predominantly in large herds on the prairies, though also found in the desert as their grazing habits are typically varied. They range from the southern states of Canada, throughout the prairie lands of central and western N. America and down into Mexico. Though only one species, this is divided into 5 sub-species as follows: Baja Californian, Sonoran, Mexican, Peninsula and the Oregon. The Pronghorn is brown and white and is aptly named from its backward curving sheathed horns with forward pointing prongs. It is commonly believed to be the 2nd fastest land animal on earth after the cheetah, reaching speeds in excess

Photo: Pronghorn © Chris Bailey

of 70 kph; it also has 320° vision and can see for up to 6.5 km. Both of these adaptations imply that its' ancestors was prey to large, fast predators that no longer stalk the plains. Wolves, coyotes and bobcats will, however, take young and weakened Pronghorns. When threatened or alarmed the white rump fluffs up, perhaps as a warning to other pronghorns. Mating occurs in the Autumn, and males produce a powerful musk exuded from glands on the side of the head. The calves are born the following Spring.

Conservation Status:

- IUCN Red List – Least Concern.
- Federal Government of Mexico – All endemic Sub-Species Endangered.
- US Fish & Wildlife Service – 2 Sub-species endangered (Mexican and Sonoran).
- Appendix 1 CITES – Mexican sub-species.

Current Status:

Colloquially named 'antelope' they are not true antelope and are unrelated to their African Plains' counterparts. They are in fact the only extant members of the family Antilocapridae of which there were once 12 genera. Once found in vast herds, their numbers declined after European settlement. Now they enjoy some protection and hunting is limited, as a result their numbers have recovered to non-critical levels. There is however a growing issue of the spread of Blue Tongue disease transmitted from sheep. They are also poor jumpers and their migrations can therefore be impeded by farmers' fences.

Image: Pronghorn © Sandra Hughes 2010

References
www.fws.org
www.natureconservancy.ca
www.arkive.org
http://animals.nationalgeographic.com
www.arizona-leisure.com

The implications, therefore, of habitat decline here, and the benefits of preserving, and indeed restoring, prairies cannot be overstated.

There are currently many organisations working hard to restore and maintain prairies across North America and Canada including the World Wildlife Fund (WWF) and the American Prairie Foundation (APF). In many cases this has involved a complete halt in farming across large tracts of land and now many reserves exist across the United States and Canada, and these areas are important for numerous prairie species. However, almost by definition, prairies are massive-scale habitats, and it is not practical to continue to take more and more land completely out of production.

Farming on Prairie Land

Landowners continue large-scale farming of prairie lands, but many are working hard to incorporate conservation within their farming practices. Soil erosion is one of the major issues for landowners on these plains. Soil erosion is covered in greater detail in Chapters 2 and 5, however, it is significant here also, as so many prairies have been turned over to monocultures such as soyabean and corn – bare soil erodes easily after harvesting. In the past, crop rotations were commonly used, and many farmers are now reverting to this practice. This is not so much going back as going forward, as modern research and technology can be applied. Using the old three-way rotation of – small grain: hay: row crop – but using more species within that rotation, for example 10 instead of 2 or 3 grains, 6 instead of 2 or 3 grassland species, and so on, plus using more perennial species such as hairy vetch (*Vicia villosa*) as a green manure, soil remains covered throughout the year and use of solar energy is maximised. Less productive, marginal areas can still be planted with other prairie species to encourage further biodiversity within this habitat. The restoration of waterways can also be facilitated using this system as less pesticide and N and P applications are required for the crops and erosion is notably reduced (Jackson and Jackson, 2002). Although many of the plants in this system have lower value yields, much of this loss would be offset by the reduction in input costs and application labour costs. This is still, of course, farming, and most of the cropping species used are not necessarily original prairie grassland species, as Laura Jackson points out that '*while these pastures may not look like prairie restoration to a botanist, they apparently do to a bobolink*'. This combination of prairie nature reserves and modified rotational cropping systems could well be the compromise that restores biodiversity to the prairies whilst continuing to feed the populace.

Switchgrass

Whilst studying farming systems on a global scale, one common theme throughout the industry becomes plain to see. That is, with the intensification of farming comes the removal of natural plant species. That may seem obvious, for instance, if you require wheat where only scrub grows it becomes necessary to remove the scrub completely and plough the land. However the environmental implications of growing a monoculture of nutrient-hungry crops over large areas of land cannot be overstated.

Grow a plant where it does not belong, and to maintain its health for harvesting, it must be protected from pests, disease and competitors by applications of chemicals; pH levels may need adjusting, often high levels of nutrients may need to be applied and watering may be required. Not to mention the implications for local wildlife when the species forming their habitat have been removed and replaced by this sanitised alien cultivar!

One of the imperatives of conservation on farmland is to reintroduce natural plant species to the land. In the southeastern states of North America, one species that has caught the attention of, not only farming conservationists, but also business entrepreneurs, is switchgrass (*Panicum virgatum*). Case Study 3.2 looks at the move towards using this American native species within farming.

Case Study 3.2: Switchgrass (*Panicum virgatum*)

Switchgrass is a tall C4 prairie grass native to North America. Historically, this species would have been part of a suite of grasses forming prairie habitat over vast areas of plain. It is perfectly adapted for poorer soils, is resistant to many plant diseases and tolerant of wide variations in rainfall, that is it can tolerate both flooding and drought. This grass can therefore be cost effective for farmers as it can be grown without large inputs of nutrients and chemicals. As it is a perennial grass it has a substantial root system where the root penetrates deep into the soil profile which helps to reduce erosion, a major conservation topic that is covered in greater detail in Chapters 2 and 5.

Switchgrass has a number of other benefits to the farmers. Yield can be twice that of other grasses, usually 2–3 tonnes/ha per annum, sometimes more in the sunnier states. Also, this grass is a significant cash crop as it can be used as biofuel, at a time when there is an ever-increasing demand for fuel crops to replace fossil fuels. This is done by using fungal and bacterial cellulases to break down the woodier parts of the plant and yielding cellulosic ethanol C_2H_5OH. Also, research is currently ongoing at Clemson University to analyse the potential of using bacteria derived from the intestinal tract of manatees for this purpose, as rates of breakdown are potentially enhanced (Henson, 2008). Of course, though many farmers are growing switchgrass, food crop production is still essential, so for many such a change in cropping systems would be inappropriate. However, research by the USDA's Native Grass Breeding Consortium is currently underway to assess the viability of making better use of the more marginal land areas on such farms by selecting switchgrasses with high N use efficiency where N levels are low. The same study is also looking at the potential use of adding legumes to the grass mix to increase N availability (www.ars.usda.gov – accessed June 2009).

Switchgrass also benefits the environment by actually sequestering high levels of atmospheric and ground carbon dioxide, CO_2, one of the main gases associated with global warming. Research indicates that after the first year of growth CO_2 sequestration can average around 0.6 tonnes per acre (0.24 tonnes/ha) per annum, though after a number of years the rate may decrease as the grasses age (Skinner and Adler, 2009).

The benefits to wildlife are also paramount. These tall grasses provide significant cover for small mammals and birds, and those farmers who have planted switchgrass have noted an increase in native wild species. We visited one such farm.

Perry Brothers Farm, Zebulon, North Carolina, USA

We met Larry Perry on a very wet morning; we had little time with him as he was on his way to a committee meeting with his colleagues in the 'Conservation Reserve Programme' (CRP). Though time was short, his infectious enthusiasm for the conservation work he is undertaking was plain to see, here was a man with a passion, keen to tell his story.

Larry's family have been farming here since the early 1800s and throughout this time they will have seen many changes to the landscape, none more so than now with increasing numbers of people moving to North Carolina and the constant threat of seemingly uncontrolled development. Larry is very active in the CRP and has collaborated with three other farmers (two in North Carolina and one in West Virginia) to place easements on their combined land consisting of a total 450 acres (180 ha). The CRP is in part responsible for advising the County Commissioners of the best land to purchase for preservation. These relatively large areas of land are being farmed sympathetically for wildlife thus maintaining the natural look of the landscape that they hold dear. At the Perry farm there are a few horses grazing pasture, but otherwise there is no livestock, the land is farmed for hay, wheat, beans and some tobacco (once one of the main cash crops in the area, but now farmed by quota under licence). Larry has introduced switchgrass to a number of his fields and informed us that, 'switchgrass is a tall dense grass which historically relies on lightening-induced burn offs. This is recreated on a three year rotation with a controlled burn'. The fields have a 4 m strip all the way round acting as a firebreak; these burns remove the thatch and kill off invasive species such as the sweet gum (*Liquidambar styraciflua*), a native American witch-hazel and a local pioneer species. Other fields have been sown with mixed annual grasses.

Using this growing system, Larry has seen an increase in small game animals such as rabbits and wild turkeys, and is hoping to see an increase in the number of Bob White quail (*Colinus virginianus*), a highly endangered species. Species Box 5.4 (Chapter 5) profiles this bird. Evidence can be seen of small mammal activity, where 'runs' are evident in the sward (Figure 3.11).

There are nest boxes around the fields for another key species, the Eastern bluebird (*Sialia sialis*), once one of North Carolina's most common songbirds.

Recently Cardiff University School of Biosciences carried out growth trials to look at the viability of switchgrass being cultivated in the UK. At the end of their trials they did not consider this a good prospect as it was out-competed by the local weed species and considerable use of herbicides was required for useful growth. This further vindicates our assertion that good husbandry in association with on-farm conservation entails the use of

Figure 3.11   Small mammal run in switchgrass. (Photo: Sarah Burchett)

native plant species, an ideal grass to grow in North Carolina, but possibly not the best option in the UK. However further tests indicate that it may yet be a useful energy crop in Europe due to its wide environmental adaptation and tolerance.

In December 2007, George W. Bush signed a new energy law mandating a 5 × increase in ethanol blending for fuels by 2022. Some of this will come from switchgrass, though what percentage that will be will depend on take-up rates by farmers over the next few years. The lifecycle of switchgrass emits 94% less CO_2 than gasoline (Schmer *et al.*, 2008). With oil-based fuels set to run out this century, and with the imperative to reduce carbon emissions into the atmosphere to reduce effect of global climate change, this puts ethanol production technology high on the agenda for the immediate future. There has been some concern that large-scale corn production (currently already being used for ethanol production) could jeopardise food security in the US, indeed the American 'bread basket' has already been reported to be under pressure, switchgrass may offer an alternative. Larry Perry points out that *'switchgrass is very forgiving, it can be grown on marginal land leaving the best land for more demanding crops'*.

Growing Switchgrass

There are many varieties of switchgrass, but they mainly fall into two categories, the lowland types and the upland types. The lowland varieties produce higher yields, though they are not fully frost hardy, therefore variety selection should be based on the environment into which it will be sown.

Although switchgrass is tolerant of varying environmental and physical conditions, it does best in medium to deep sandy and loamy soils, and does not grow as well in heavy soils, therefore if it is to be grown specifically for biomass production, then heavy clay soils would not be suitable. It is also tolerant of a fairly wide pH range, though optimal pH is about 6.5. It does well in poorer soils, and though the addition of N may increase early yields, it is unnecessary as harvesting is not advisable in the first year, and increased levels of N will also encourage competitors.

Switchgrass seeds are very fine, so good soil preparation is essential to ensure a good even coverage; a windy day would not be advisable! The application rate is approximately 8–10 kg/ha, sown after all chance of frost has gone as seedlings must develop to hardiness, at a depth of around 1 cm. Weeding out competitors, where possible, in the first year will be beneficial to establishment of good coverage, however, it is a tough, rhizomal grass so success rate is high and chemical weed killers may cause damage to the grass as well as being non-conducive to the on-farm conservation philosophy. Nutrient applications in following years are not really necessary as the stubbles are left in after cutting and nutrients will leach back from these into the soil, however, if high yields and multiple cuts are required for ethanol production, the addition of N may be advisable at a low rate of around 60 k/ha.

Summary

Switchgrass grows well in its native land of America. It is high yielding, requires little by way of additions, can be used as forage and in biofuels and is recommended in on-farm conservation as good habitat for a number of small mammals, birds and invertebrates. There have been some moves towards growing this grass elsewhere in the world for ethanol production though its role in on-farm conservation is not necessarily applicable outside of the US.

3.4.2   Grasslands of the British Isles

On a geographical scale the British Isles is relatively small, however with respect to geology, climate, soil type, habitat type and consequently biodiversity, it is comparable with much larger-scale temperate regions across the world. There are mountains, uplands and lowlands, farmland and forest, lakes, river systems and maritime habitats with numerous islands. Acidic peaty moorland to alkaline chalk downland, high rainfall areas to low rainfall areas and a climate influenced by the gulf stream that sweeps around the west coast from the Atlantic arriving first in Cornwall and Ireland, keeping the region temperate despite being on the same latitude as Siberia. Once heavily forested, various types of grassland are now the most predominant feature in the landscape (Table 3.7).

3.4.3   The uplands

The uplands of the UK are defined by the 'moorland line', which describes land that is predominantly semi-natural upland vegetation and used for livestock grazing and commercial softwood plantation, in Scotland game shooting for deer and grouse is also

Table 3.7 Grassland habitats

Habitat	Characteristics
Acid grassland	Dominated by *Agrostis* grasses and heath species on lime deficient and acid soils.
Dwarf scrub heath	Well drained, nutrient poor acid soils. Cover of heath species greater than 25%.
Calcareous grassland	Dominated by grasses, but high diversity of forbs and orchid species. Occurs on chalk, limestone or other basic rich rock. Characterised by well drained nutrient poor soils.
Fen, marsh and swamp	Characterised by permanently or seasonally waterlogged soils. Can be composed of peat, peaty type soils or mineral soils.
Montane	Grasslands and associated vegetation that occurs above the tree line. Includes low growing dwarf shrub heath, snow-bed communities, sedge, rush and moss heaths.
Inland rock	Lower altitude habitats that occur on natural and artificially exposed rock. Including tors, inland cliffs, caves, screes and limestone pavement.
Bracken	Stands of vegetation, greater than 0.25 ha in area dominated by bracken during the height of growing season.

a significant land use. The moorland line is established around the 300 m contours, using Ordnance Survey, mapped features such as walls, tracks or streams (Defra, 2006). In England the moorland line includes areas such as Dartmoor, the Peak District, the Lake District, the Forest of Bowland, North Pennines and the North Yorkshire Moors. In Wales the moorland line encompasses much of Snowdonia, the Brecon Beacons and the spine running between them. In Scotland the line encompasses many more areas including much of the Highlands and Islands, Shetland but not much of Orkney, and major blocks of Scottish lowland hills. In the wider European community many agricultural areas fall above the 300 m contour and as such are upland areas and experience significant handicaps to successful and sustainable farming practice. This has been recognised by the EU government which has developed a set of policies to aid farmers, the most significant being the less favoured area (LFA) designation of 1975, under Council Directive 75/268/EEC. Originally there were two categories of LFA, Mountain areas and Hill Farming areas, which were identified using a mixture of physical and socio-economic data. The original designations are now incorporated within the Council Regulation 1257/1999.

The rationale for LFA designation is the recognition that across Europe many of these areas were being abandoned or were under imminent threat of abandonment because of shifting socio-economic factors and increased industrialisation of agriculture. Conservationist and many policy makers in the EU community acknowledged that such upland areas contained many threatened habitats; such as heather moorland and Rhôs pasture, and numerous species of conservation concern, including species such as the golden plover (*Pluvialis apricaria*), (Species Box 3.3) black grouse (*Tetrao tetrix*), the red grouse (*Lagopus lagopus scotica*) and the lapwing

Species Box 3.3: Golden Plover (*Pluvialis apricaria*)

Profile:

A small wading bird, though fairly large for a plover (27 cm), the golden plover is reliant on upland moorland habitat during the breeding season. It has striking gold and black summer plumage and a distinctive light coloured 'V' across back and wings when in flight. They nest in a shallow scrape lined with lichen and heather, have four eggs and raise only one brood in late spring and early summer. In the British Isles they breed on moorlands in Scotland, Wales, NW Ireland, Yorkshire and Devon. In the winter they migrate to lowlands, mudflats and estuaries in southern Britain and Ireland, parts of Europe and Northern Africa, often in large groups in the company of lapwings. They feed mainly on worms, insects and sometimes berries in the summer, and on crustacea and worms on their winter feeding grounds.

Image: Golden Plover © Sandra Hughes

Conservation Status:

- IUCN Red List – Least Concern.
- Dartmoor species action plan.

Current Status:

Wide-ranging and abundant, not considered currently under threat nationally in Britain, however numbers on Dartmoor National Park are critically low and in decline probably due to changes in habitat management strategies which affect blanket bog habitat.

References

Elphick, J. and Woodward, J. (2009) *RSPB Pocket Birds of Britain and Europe*, 2nd edn, Dorling Kindersley Ltd (DK), London.

Dartmoor Biodiversity Steering Group (2001) *Action for Wildlife – The Dartmoor Biodiversity Action Plan*, English Nature.

(*Vanellus vanellus*) (Species Box 3.4). Abandonment threatens these habitats and species within them because grasslands require constant management, which in the uplands comes from grazing animals. However, during the 1970s and 1980s many upland areas in the UK were overgrazed, this was the result of headage payments made to farmers, under CAP (Chapter 1) for each animal raised and consequently this led to increased stocking density. In the UK uplands this increase in livestock adversely affected the vegetation structure of heather moorlands leading to a decline in the vigour of plants and stand density. Furthermore many palatable grasses (the *Fescues* and *Agrostis* species) were overgrazed, leading to an increase in

Species Box 3.4: Lapwing (*Vanellus vanellus*) – Also Known as the Peewit

Profile:

Wide-ranging wader found across Britain and Ireland, Europe, Asia and Africa, it is the largest of the European plovers (30 cm). Summer habitat is open grassland, especially arable land and boggy heath moorland. Nests in shallow hollows on the ground, lined with grass and lays three or four eggs in a single brood, mid-March to mid-July. A highly distinctive bird, with notable long wispy tuft on the head, a distinctive flight that appears sluggish, though the male has an impressive display flight when courting. In the winter they migrate to lowland fields, mudflats and saltmarsh and are often seen in large numbers with golden plovers and gulls. They eat mainly soil-dwelling invertebrates.

Image: Lapwing © Sandra Hughes.

Conservation Status:

- IUCN Red List – Least Concern 2009.
- RSPB Red Status.

Current Status:

Internationally not considered under threat, however in Britain numbers have dwindled significantly possibly due to changes in agricultural practice.

References

Winspear, R. (ed.) (2007) *RSPB Farm Wildlife Handbook*, RSPB, Sandy, UK

Elphick, J. and Woodward, J. (2009) *RSPB Pocket Birds of Britain and Europe*, 2nd edn, Dorling Kindersley Ltd (DK), London

Rosair, D. and Cottridge, D. (1995) *Photographic Guide to the Waders of the World*, Chartwell Books, Inc.

unpalatable species such as purple moor grass (*Molinia caerulea*) and mat grass (*Nardus stricta*). During this period of heavy grazing the damage inflicted on natural vegetation led to an increase in other unpalatable species such as bracken (*Pteridium aquilinum*) and gorse (*Ulex gallii and U. europaeus*), which, although taken by livestock, are not preferred dietary items. By the end of the 1980s it was recognised that too many animals were being grazed on the uplands and in the UK schemes were established to encourage removal of grazing animals from many upland areas, in particular Environmentally Sensitive Area Payments and Country Side Stewardship Schemes. However, as in the past the application of rigorously applied theoretical stocking rates has brought about dramatic and disastrous changes in the composition of upland vegetation. This well meaning but poorly applied philosophy of de-stocking has resulted in many upland areas being completely or largely dominated by dense stands of bracken, gorse and purple moor grass. In the wetter areas that were once Rhôs pasture or valley mires, bogs and other wet flushes, removal of livestock is leading to a succession from mire vegetation to Willow Carr woodland. What is clear is that many policy makers and theoretical ecologists are far removed from the real world of the uplands. One must remember that grazing created the uplands, which initially were based on low stocking density and minimal artificial inputs. The following case study is based on Dartmoor, in southwest England, and uses a small family farm to illustrate the challenges facing many upland farmers across the European Community.

Case Study 3.3: UK Uplands

Dartmoor

Although man made and, quite candidly, something of an industrial wasteland created by mining, quarrying and grazing, Dartmoor is designated as a natural area (English Nature, 2001). This is quite understandable as many semi-natural landscapes and habitats prevail on the moor, supporting numerous rare and endangered species (Map 3.2; Table 3.8). Granite is the dominant parent rock and Dartmoor is described geologically as intruded granite massive, surrounded by carboniferous sandstones, shales and grits. Towards the southwestern end of the moor the parent material is dominated by upper and middle Devonian slates and limestone (English Nature, 2001).

The underlying geology, altitude and cold wet prevailing climate has influenced the development of plant communities and consequently Dartmoor contains quite a rare habitat type in southern England, matched in character and habitat heterogeneity by both Exmoor in Devon and Bodmin Moor in Cornwall. Consequently, many of the habitats on Dartmoor are listed in the UK BAP plan such as Rhôs pasture, culm grassland, moorland heaths, blanket bog and valley mire, upland oakwoods, parkland and wet woodlands and all are listed as priority habitats. A UK priority habitat is a habitat that is internationally recognised as an important feeding and/or breeding ground for many priority species and, as such, the UK has international obligations in maintaining and improving the extent and quality of such habitats (Dartmoor National Park Authority, 2001). Dartmoor also has a number of designated sites in particular Sites of Special Scientific Interests (SSSIs), which include geological features, rivers, woodlands, mires, moorland heaths and grass moor. There are 57 SSSIs on Dartmoor (English Nature, 2001). Finally, Dartmoor is also designated as a National Park.

Managing the habitats on Dartmoor is a complex multi-agency effort that includes English Nature now Natural England, Dartmoor National Park Authority, Dartmoor Preservation Association, DEFRA and the Environment Agency. These agencies do not work in isolation but endeavour to develop working relationships with many landowners (farmers) on Dartmoor in an effort to maintain and enhance habitats and species diversity. Sometimes these partnerships have mutually beneficial outcomes, but sadly this is not always the case and like much of the UK uplands there are some very poor management practices being applied on Dartmoor, typically as a result of poor developments in land use policy such as over-zealous de-stocking and the removal of winter grazing. The management of Dartmoor also includes a number of active commons where many moorland farmers have grazing rights.

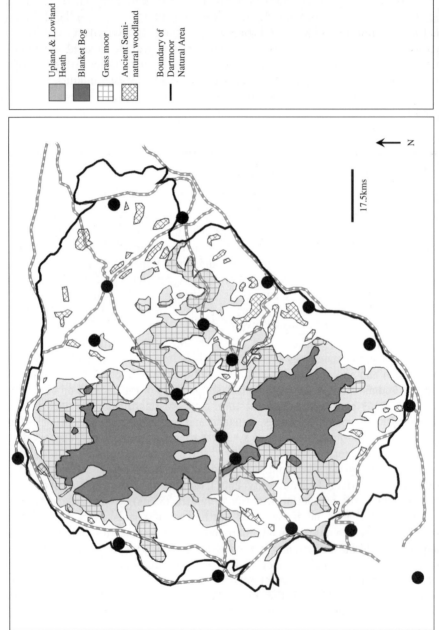

Map 3.2 Habitats observed in the Dartmoor Natural Area. (Adapted from Dartmoor National Park Authority, 2001)

Table 3.8 Species list for Dartmoor

Main habitats	Component habitats	Key species	
		Scientific Name	Common name
Moorland	Upland heath	*Alauda arvensis*	Skylark
		Boloria euphrosyne	Pear Bordered Fritillary
		Calluna vulgaris	Heather (Ling)
		Lagopus lagopus	Red Grouse
		Pluvials apricaria	Golden Plover
		Turdus torquatus	Ring Ouzel
	Lowland heath	*Alauda arvensis*	Skylark
		Boloria euphrosyne	Pear Bordered Fritillary
		Calluna vulgaris	Heather (Ling)
		Euphrasia vigursii	Vigur's Eyebright
	Grass moor	*Alauda arvensis*	Skylark
		Argynnis adippe	High Brown Fritillary
		Calluna vulgaris	Heather (Ling)
	Blanket bog	*Calidris alpina*	Dunlin
		Lagopus lagopus	Red Grouse
		Pluvials apricaria	Golden Plover
	Raised mire	*Eurodryas aurinia*	Marsh Fritillary
	Valley mire	*Coenagrion mercuriale*	Southern Damselfly
		Eristalis cryptarum	Bog Hoverfly
		Eurodryas aurinia	Marsh Fritillary
		Hammarbya paludosa	Bog Orchid
		Numenius arquata	Curlew
		Orthetrum coerulescens	Keeled Skimmer Dragonfly
		Spiranthes romanzoffiana	Irish Lady's Tresses
Woodland	Upland oak	*Buteo buteo*	Buzzard
		Carabus intricatus	Blue Ground Beetle
		Graphina pauciloculata	Lichen
		Muscardinu avellanarius	Dormouse
		Narcissus pseudonarcissus	Wild Daffodil
		Usnea articulata	String of Sausages Lichen
	Wet woodlands	*Usnea articulata*	String of Sausages Lichen
	Plantation	*Buteo buteo*	Buzzard
Freshwater	Torrent rivers	*Lutra lutra*	Otter
		Salmo sala	Salmon
	Streams	*Lutra lutra*	Otter
		Salmo sala	Salmon
	Reservoirs	*Lutra lutra*	Otter
	Ponds	–	–

(continued overleaf)

Table 3.8   (*continued*)

Main habitats	Component habitats	Key species	
		Scientific Name	Common name
Farmland	Rhos pasture	*Emberiza cirlus*	Cirl Bunting
		Muscardinu avellanarius	Dormouse
		Calidris alpina	Dunlin
		Coenagrion mercuriale	Southern Damselfly
		Eristalis cryptarum	Bog Hoverfly
		Eurodryas aurinia	Marsh Fritillary
		Hemaris tityus	Narrow Bordered Bee Hawkmoth
	Haymeadows and species rich grassland	*Buteo buteo*	Buzzard
		Lullula arborea	Woodlark
		Dianthus armeria	Deptford Pink
		Maculinea arion	Large Blue Butterfly
		Platanthere chlorantha	Greater Butterfly Orchid
		Rhinolophus ferrumequinum	Greater Horseshoe Bat
Field boundaries and features	Walls	*Alauda arvensis*	Skylark
		Rhinolophus ferrumequinum	Greater Horseshoe Bat
	Hedges	*Rhinolophus ferrumequinum*	Greater Horseshoe Bat
		Platanthere chlorantha	Greater Butterfly Orchid
	Isolated and vetran trees	*Usnea articulata*	String of Sausages Lichen
Geology	Quarries, rocky outcrops, caves and mines	*Rhinolophus ferrumequinum*	Greater Horseshoe Bat
		Turdus torquatus	Ring Ouzel
		Dianthus armeria	Deptford Pink
		Hypericum linarifolium	Flax Leaved St Johns Wort

Drywall Farm Widecombe-in-the-Moor

Drywall Farm is a typical, small, family run, Dartmoor farm, extending to some 32 ha (80 acres) located in a protected wooded valley southwest of Hameldown (Map 3.3). Hameldown (Figure 3.12) rises to 532 m and is a large expanse of upland heath and grass moor to the east of Widecombe-in-the-Moor. The farm has been in Sue's family for many generations and such ancestry is commonplace on small farms in Devon and Cornwall, indeed the herd of South Devons that Sue runs can be traced back to the mid 1880s, when they were first introduced to the farm. This long ancestry has helped to reduce the susceptibility of cows and calves to tick borne diseases such as red water fever and tick fever. Many upland farmers develop their livestock by selecting breeding animals from neighbouring farms, because animals bred and raised in upland scenarios have developed a natural resistance to many tick borne diseases.

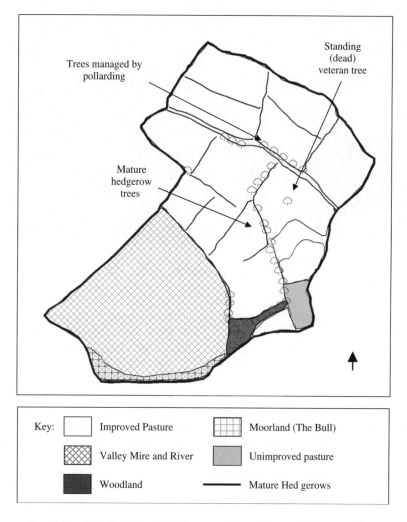

Map 3.3 Distribution of habitats and grasslands on Drywall Farm

Although the farm lies in a protected wooded valley, this should not be mistaken as implying the climate is benign and the land is all good agricultural land. The truth differs with, about 15 ha given over to the production of grazing and silage production, the remainder being occupied by semi-natural grasslands and a river valley (Map 3.4). The production side of the farm is based on a low input and low output philosophy which is achieved by using two traditional breeds of grazers; the South Devon cow and the Llyen sheep. The South Devon (Figure 3.13) is an excellent, docile beef suckler cow and performs

Figure 3.12　Hameldown, Dartmoor taken from the air. (Photo: Stephen Burchett)

exceptionally well on the limited grazing available at Drywall Farm. Their docile nature is essential for Sue, as she often has to manage and handle the herd on her own. Such endeavours are typical on small family farms. Another key aspect of the South Devon is it has a ready market in Devon, where a number of regional butchers actively seek out the breed because of its exceptional meat quality. The Llyen ewe (Figure 3.14) is again another excellent breed for a low input system, the Llyen is a white faced, hornless, medium sized sheep, the ewes weighing about 75 kg when mature. The breed originates from the Llyen peninsular in North Wales and thus is inherently hardy, performing well on poor grasses and upland habitats. The fleece is very fine, with no kemp and produces very fine wool that is used in hand knitting, hosiery and dress fabrics. The ewe will produce two lambs and is an exceptional mother so that lamb mortality is very low and the ewe is noted for her ease of lambing. Finally, the meat from the butchers' lambs is lean and good quality with a high grading out percentage thus maintaining farm income.

Using these two traditional breeds Sue has managed to reduce her farm nutrient inputs from 6 to 1 tonnes across the farm per year. This is a significant saving in expenditure and consequently also a reduction in the concentration of inorganic nutrients leaching into neighbouring watercourses. Both breeds can be grazed on the improved pasture and the rough semi-natural grassland (Map 3.4). In total Sue has 30 South Devon cows and 12 calves and 40 Llyen ewes and lambs at foot.

Commercially this is a very small herd and flock, but by adapting the farm business to a low input system and choosing breeds that can perform well in a low input system Sue has helped to reduce costs and keep the farm viable.

Figure 3.13   South Devon cows. (Photo: Stephen Burchett)

Figure 3.14   Llyen ewes. (Photo: Stephen Burchett)

Conservation Work at Drywall Farm

The farm has two agri-environment schemes, ELS and HLS, which support conservation work on an area of semi-natural grassland called the Bull (Map 3.3) and a stream and adjacent bank side. The Bull has a number of habitats including upland heath and grass moor and valley mire, which effectively runs parallel with the stream (Map 3.3). The mire is an excellent stronghold for the devils-bit scabious (*Succisa pratensis*), a favoured food plant for the pearl-bordered fritillary (*Boloria euphyrosyne*) and marsh fritillary (*Eurodryas aurinia*) butterflies which are known to use the mire. Sue manages both these habitats using a mixture of cutting and light grazing with six Dartmoor ponies. Sue avoids using the Llyen ewes because both habitats have a significant amount of brambles (*Rubus friticosus*), which catch in the fleece reducing its commercial value. The HLS scheme imposes a closed period for grazing from 1 September to May 31 and it has to be noted that this is leading to a significant encroachment of gorse and bracken (Figure 3.15). Sue is struggling to manage the growth of these encroaching species and noted that areas cleared in 2004 are now fully encroached again. This anecdotal evidence clearly illustrates how some conservation policies supported by government grants are failing to meet their conservation targets and demonstrates that further scientific research and applied conservation practice needs to be evaluated to improve policy decisions and management recommendations (Bangor University Dr Andrew Pullin, Conservation evidence unit).

Figure 3.15   Bracken encroachment on Dartmoor. (Photo: Stephen Burchett)

Managing the stream and bankside has been a key feature of the conservation work in 2009, and Sue has coppiced some trees to open up the canopy around riffle beds and encouraged shade around the deep pools to enhance aquatic diversity. This had been done in conjunction with coppicing willow and cutting gorse alongside the stream bank and adjacent valley mire to improve stand vigour and regeneration of the devils-bit scabious (Figure 3.16).

On-farm conservation is not limited to the stream and Bull, the hedges at Drywall are in excellent condition, with rotational cutting resulting in a diverse hedgerow community of tall dense hedges. These hedges support numerous foraging species including the pipistrelle bat (*Pipistrellus pipistrellus*), the long eared bat (*Plecotus auritus*) and the greater horseshoe bat (*Rhinolophus ferrumequinum*). The old barns that were once used for livestock were

Figure 3.16   Devil's bit scabious. (Photo: Stephen Burchett)

Figure 3.17   Bat vents on roof. (Photo: Stephen Burchett)

repaired and renovated under CSS to provide roosts for some of these bat species by incorporating small bat size vents into the tiled roof (Figure 3.17). The farm also has a number of mature trees, oak and beech, and in 2008 a mature oak estimated to be over 800 years old died, but instead of cutting this tree down Sue has left it standing to act as a habitat in its own right (Figure 3.18). Finally, many of the mature trees in the

Figure 3.18   Ancient oak, left to stand after dying as wildlife refuge. (Photo: Stephen Burchett)

hedgerows are managed by pollarding. Pollarding is where trees are cut at approximately 4 m above the ground every 15 years, this activity helps to maintain longevity of the trees and consequently improves their age structure, which helps to enhance landscape character and provide a range of structural habitats for local avian fauna.

All this valuable conservation work is integrated into the daily routine of the farm and Sue is an important community member volunteering her services during swaling events on Widecombe common. The skills and knowledge Sue has amassed over her lifetime at Drywall is invaluable and quite irreplaceable. However, due to the small size of the farm, it is quite likely that when Sue eventually retires from farming all this knowledge and experience will be lost. It is important that the UK government and indeed the wider EU government recognises these problems, and in some respects they have done so with the implementation of the LFA Status, but in other ways governments implement policy that could push many more upland farmers out of business. It is essential that Europe continues to support small-scale farms and governments continue to invest in educating the next generation of hill farmers, sadly again an area neglected by successive UK governments.

In the British Isles many grassland meadows were once rich in abundant flower species. These flower meadows evolved alongside low-level farming practice. In modern times intensive farming and stock grazing on improved grasslands all but wiped these wildlife havens out. In recent years the conservation value of such meadows is recognised and conservation grazing practices are increasingly being implemented to both preserve and reintroduce these habitats.

3.4.4   BurrenLIFE project

One innovative project is the BurrenLIFE Project. Situated in an area called The Burren in County Clare, south of Galway in the west of Ireland, this project is particularly pertinent to conservation farming and biodiversity as it boasts no less than 12 different habitat types, listed below, covering an area little over 16 000 ha and made up of: the East Burren Complex; the Black Head Poulsallagh Complex and the Moneen Mountains – all protected under the EU Habitats Directive. The project is 75% funded by the EU LIFE Project, and run by the National Parks and Wildlife Service (NPWS), Teagasc – the Agricultural and Food Authority and the Burren Irish Farmers Association plus many supporting bodies, and operates on the basis that one of the major factors involved in shaping this land was farming. As such, the best means to conserve and maintain these habitats is to farm in a sustainable and traditional way. Conservation grazing is one of the main methods used to maintain the grassland habitats. A number of rare and unusual species can be found in the species-rich grasslands, including many orchids such as the frog orchid (*Coeloglossum viride*), eyebrights (*Euphrasia spp.*) and the spring gentian (*Gentiana verna*).

The Burren Habitats

Limestone Pavement; Dry Calcareous Grassland; Neutral Grassland; Wet Grassland; Improved Agricultural Grassland; Limestone Heath; Scrub and Woodland; Turloughs; Calcareous Springs; Lakes; Fens and Blanket Bog.

3.4.5   The Mynydd Mawr Marsh fritillary project

The Mynydd Mawr Marsh Fritillary Project is another innovative initiative that coordinates landowners in an area of Carmarthenshire in Wales to carry out low-level stock grazing, with the prime objective of protecting and maintaining habitat for the marsh fritillary butterfly (*Euphydryas aurinia*). The butterfly requires damp tussocky grassland, known as Rhôs pasture, with an abundance of devil's-bit scabious (*Succisa pratensis*). Scabious is the only food plant of the caterpillars and is therefore essential to the life cycle of the butterfly. The project has been responsible for linking land areas with this habitat, thus ensuring its long-term success. They have also coordinated the grazing regime by networking landowners with no grazing stock to those with too much (relative to the grazing levels required), plus the project has helped with practical functions such as fence building and scrub clearance. Grazing is an important factor in maintaining this habitat. If overgrazed no shelter is provided for the butterflies, if left ungrazed, the grasses outcompete the flowers in the meadow. At Mynydd Mawr Welsh Black Cattle and ponies are used for this function, but not sheep as they graze too low and eat the flowers. A well developed flower meadow habitat such as those in Wales also support numerous other plants including a number of orchid species, ragged robin (*Lychnis flos-cuculi*) and whorled caraway (*Carum verticulatum*); birds including barn owls (*Tyto alba*) and grasshopper warblers (*Locustella naevia*); and other animals including dormice and voles, (Butterfly Conservation).

Chalk Grasslands are rare habitats of international concern and are often classed as SSSI. They are distributed only in northwest Europe and the remaining chalk grasslands are extremely rare. In the UK there are around 33–41 000 ha of calcareous grassland and the majority of these habitats are found within the south and east where the climate is warmer and milder (UKBAP, 1998). The South Downs region is the largest area of calcareous grassland in the UK. The massive decline in this type of habitat is thought to be due to a reduction in sheep farming.

These habitats are rich in biodiversity but can differ depending on their location and climatic differences. These rare grasslands provide a habitat for many nationally scarce species of wild meadow flowers, insects and birds (Beldon *et al.*, 2000). Unfortunately they have severely declined within the last 50 years, this has been caused by various factors such as agricultural intensification, farm specialisation towards growing crops, over-grazing causing damage to sward, infrastructure developments, recreation pressures, invasive species such as *Cotoneaster*, along with atmospheric pollution and climate change (UKBAP, 1998).

Grazing has been a major form of land management for several 100 years and has thus helped create many of today's British landscapes, including chalk grasslands. Conservation grazing – grazing that meets and aids nature conservation purposes, is vital as livestock removes vegetation biomass and thus allows less competitive species to grow and to increase in dominance, whilst trampling can create areas to allow seed regeneration (GAP, 2007).

Park Gate Down is an example of chalk grassland. Situated on a downland site near Canterbury in east Kent, it extends to 8 ha of grassland and contains 12 different species of orchid, including the rare monkey orchid (*Orchis simia*), the lady orchid (*Orchis purpurea*), the late spider orchid (*Ophrys fuciflora*) and the bee orchid (*Ophrys apifera*) which is featured in Species Box 3.5. These orchids flower in May and June. The site is managed by the Kent Wildlife Trust and attracts national media interest having been featured on several television wildlife programmes. Conservation grazing, using Konick ponies and Highland cattle, is carried out from September to December and is an essential part of the reserve's management strategy. Overgrazing would damage the precious plants that are fundamental to the ecosystem, but equally fundamental is the low-level grazing that maintains the grass sward at an optimum level to allow for flower seeding and growth. Maintenance of the habitat for orchids in turn supports insects such as the Chalkhill blue butterfly (*Polymmatus coridon*) and birds such as the whitethroat (*Sylvia communis*) and the yellow hammer (*Emberiza citrinella*).

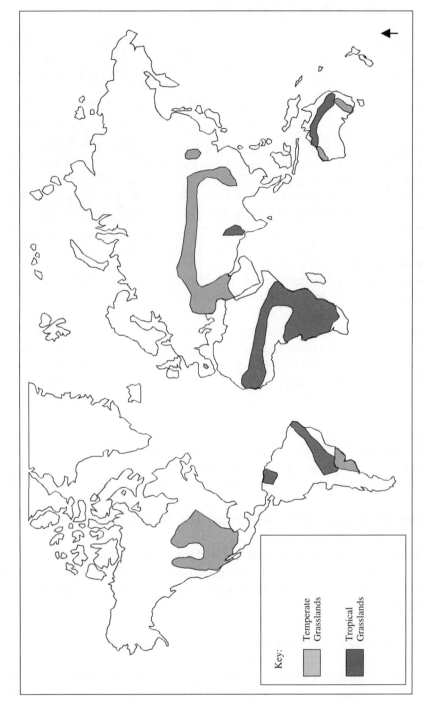

Map 3.4 Distribution of temperate and tropical grasslands. (Adapted from Sadava *et al.*, 2006). Extent of Prairie Grasslands in USA. (Adapted from McCracken Peck, 1990)

Species Box 3.5: Bee Orchid (*Ophrys apifera*)

Profile:

Flowering period June to July in the UK. The bee orchid nearly always self-pollinates, however, occasionally pollination by bees of the genera *Andrena* and *Eucera* occurs (Lang, 2004). Likes a wide range of suitable habitats including chalk, clay or calcareous sand, grasslands, sand dunes, scrub, roadside verges and industrial waste ground where there is a base-rich substrate due to weathering. Prefers well-drained soils though can exist in damp areas (Lang, 2004). The orchid lives symbiotically with a *Mycorrhizal* fungus, and though common in southern Europe, is found as far north as northern UK and is the county flower of Bedfordshire.

Bee Orchid: Photo © Stephen Burchett

Conservation status:

Protected under Wildlife (Northern Ireland) Order Schedule 8 (Lang, 2004). Though not uncommon, all orchids in Britain enjoy limited conservation protection.

Current Status:

Common in the south and east of England, less common in the southwest. Most often found on the north and south coasts of Wales, and the central counties in Ireland. Formally believed to be extinct in Scotland, until found in Ayrshire in 2003 (Lang, 2004).

References

Lang, D. (2004) *Britain's Orchids*, WildGuides, Ltd, Hampshire.
Rose, F. (2006) *The Wildflower Key*, Penguin Group, London.

3.5 Grassland ecosystems around the world

As mentioned, grasslands can be loosely divided into tropical and temperate grasslands. We say loosely because even under these headings habitat diversity is substantial. Map 3.4 illustrates the main geographical areas where tropical and temperate grasslands are found.

3.5.1 Tropical grasslands (savannas)

These are found lying to the north and south of the equatorial rainforests, usually between 5 and 15 °N and 5 and 15 °S. They cover vast areas, and form a wide array of distinctive habitat types, most featuring scattered and more densely arranged trees and scrub. Though grasslands by definition, savannas are often distinguished by the main tree species that forms the light canopy above, for example the African acacia savannas or Baobab savannas, and these grasslands support the world's greatest diversity of ungulates and other large mammals (Sodhi, Brook and Bradshaw, 2007). Tropical savannas, however, are under ongoing threat from increasing agriculture, urbanisation and desertification.

Savanna is found from the temporary floodplains of the Orinoco Basin and the Brazilian Campos and the Pampas of South America, plus areas of Mexico to areas in Western India and Northern Australasia. However, most is to be found on the African continent. Map 3.4 shows the grasslands of the world. It should be noted that some of the montane alpine grasslands are found in tropical latitudes forming cool-climate 'islands' in tropical regions. These, however, are exceptions to the rule and not classified as savanna. The African savanna grasslands are predominated by C4 grass species, though in some of the wetter areas C3 grass species become more abundant. They are mainly subjected to two seasons, the long dry season and the shorter wet season. Periodic burning, ignited by lightening, towards the end of the dry season helps to maintain the health of the grasslands and its inhabitants, as the new fresh growth is more attractive to grazing animals (Trollope and Trollope, 2004). The growth of scrub is limited and nutrients are released into the soil, which often tends to be porous and have only a very thin layer of humus (www.defenders.org – accessed October 2009). The benefits of these fires are well recognised by conservationists and landowners alike and are used as a management tool both to maintain grasslands for wildlife and for livestock production and game ranching (Trollope and Trollope, 2004).

Increasingly agriculture seems very much at odds with wildlife and environmental conservation in these types of habitat; indeed the most effective conservation strategy is to designate large areas as parks where little or no farming is permitted. Planting crops on migratory routes of the large game herds is destined for disaster, the crops are crushed and eaten by the game and livestock are taken by the predators that inevitably follow the herds. There are, however, ways to surmount these issues, for instance in Tanzania, near Mount Kilimanjaro 10 different communities have combined to set aside portions of their land to form a 250 000 acre (100 000 ha) wildlife management area, along with set-aside land from private landowners, migratory routes of large game herds can be protected (www.nature.org – accessed November 2009).

It should be noted, however, that not all of the people who live on the savanna plains have sufficient resources to designate large areas of land to such schemes and these people need to make a living from their land. There are a number of schemes in place that help to ameliorate the impact of such smaller-scale farming on the environment in such settled areas.

Lewa wildlife conservancy

One such example is the Mutunyi Irrigation Scheme in Kenya. Six hundred families live in a settlement bordering the Lewa Wildlife Conservancy Park and draw their irrigation water from the Rugusu River. The scheme devised a plan to convert the existing furrow irrigation system to a piped gravity system for the homes and combine it with an in-field sprinkler system for the crops. This increased efficiency, delivering the water that was required to the fields where it was needed. Water need not be wasted and no more water than was required was drawn from the river. It is 75% funded by the Community Development Trust Fund (CDTF) and the system came into full use in 2005 and consists of a reinforced concrete weir on the Rugusu River, a sedimentation tank, a break pressure tank on the parkland and a series of pipelines to the houses and the fields stock (www.lewa.org – accessed November 2009).

In areas where water availability can be precarious, this type of scheme benefits both the community and the wider environment. Further to the irrigation scheme, since 2005, communities bordering the Lewa Wildlife Conservancy, graze their stock on areas of the parkland in a controlled scheme that reduces the high biomass of moribund grasses that are not generally favoured by the wild game herds. Controlled burns are used as a management tool here, but conservation grazing has proven to be invaluable. The livestock reduces biomass of undesirable grasses such as *Pennisetum* species by grazing and trampling, without impacting on the small plains game, invertebrates and reptiles. The large game herds benefit from the improved habitat, and the local communities benefit by increasing the grazing range for their stock (www.lewa.org – accessed November 2009).

3.6   Temperate grasslands

So far we have discussed the chalk grasslands of the British Isles and the prairie grassland and switchgrass initiatives of the US, but there are numerous other temperate grassland types, including the steppes found in Russia and other regions, the pampas of South America, the South African Veldt, the Canterbury Plains of New Zealand and montane grasslands found globally. Temperate grasslands are found in areas where the temperature range is extreme, that is with cold winters well below freezing and warm to hot summers up to above 30 °C.

There are a number of ways in which temperate grasslands have been formed. Natural temperate grasslands are hard to define and many have argued that none are completely without anthropogenic influence. However, small pockets of grassland have evolved to be tolerant of soils containing high levels of minerals such as the serpentine soils, here only a limited number of other, highly specialised plants can survive, but grasses thrive. These areas are found globally in small, but fairly frequent pockets, for instance in parts of the US, especially California, at Cape Reinga at the very northern tip of New Zealand, the Lizard Peninsula in Cornwall, UK, parts of Turkey and in some places in the Alps, to name but a few. As discussed earlier it is also probable that much of the prairie grasslands, especially the short-grass prairies of the western regions of the US are naturally formed. A further candidate would be the montane grasslands that exist at altitudes above the treeline in mountainous regions around the world; many such grasslands are surely naturally formed.

The majority of temperate grasslands, however, are semi-natural. Semi-natural temperate grasslands are those where human influence has played a part in their formation at some point in their history, usually through large-scale deforestation when humans started to settle in communities and agriculture became a way of life. Where these lands were left then to their own devices, some reverted to forest, but many developed into the 'unimproved' grasslands that we see today through natural seeding of grass species and forbs, these were then maintained in this state by low level grazing, by periodic burning by wildfire or just because the nutrient levels in the soil were too poor to support regrowth of the forests – the nutrients in the system having been removed as a consequence of deforestation. Semi-natural grasslands are therefore regions that are not fully wild, but that do support wild populations of plants and animals within the ecosystem.

Cultivated or 'improved' grasslands contain few species and are those that are subject to regular management to a greater or lesser degree, perhaps for stock, or sometimes for recreation purposes such as golf, racing or football.

Here we highlight three examples from numerous conservation initiatives within temperate grassland farming, starting with a small case study of a farm in New Hampshire, USA where small areas of semi-natural and improved grassland form little oases for wildlife within a predominantly wooded region.

Case Study 3.4: Sheltering Rock-Tuckaway Farm

Sheltering Rock-Tuckaway Farm is a combination of farms owned by Chuck and Laurel Cox and their family in the state of New Hampshire, USA. New Hampshire is a largely forested state, around 85% of the land is covered by trees. However, the Cox's farm is predominately grassland. The land covering approximately 100 ha (255 acres), has been in the Cox family for many generations and they are actively working to restore and maintain their land for future generations. To this end it has been put into easement. Showing us around their farm, Chuck's enthusiasm for the work they do was second to none, actively encouraging visitors to walk through the lanes through the 'rights of trespass', a vestige from the UK found only in New England and helped by small tax breaks allowed for permitting open access. Of the 100 ha, 18 ha are grassland, there are a number of sizable ponds, many vernal pools, plus the Oyster River runs through the land, and the rest is forest made up largely of white pine, oak, maple and hickory, Map 3.5 shows the main features of the Cox's farm.

Grassland only covers approximately 4% of the land in New Hampshire and needs to be actively maintained to prevent natural forest succession. This is important for local wildlife; in excess of 70 species are reliant on this grassland habitat for their survival – both for feeding and for breeding. Most of this grassland is agricultural, especially hay meadows and is made up of a mixture of native and exotic species plus mainly native forbs. Maintenance of these grassland habitats is by periodic burning, by grazing, mowing and by on-farm rotations between grasses and crops (Clyde, 2008).

All farms have to make a living for the landowner and incorporating habitat maintenance and wildlife management with making a profit can be challenging. The Cox family has risen to this challenge. There is 45 acres (18 ha) grassland on the farm of which 30 acres (12 ha)

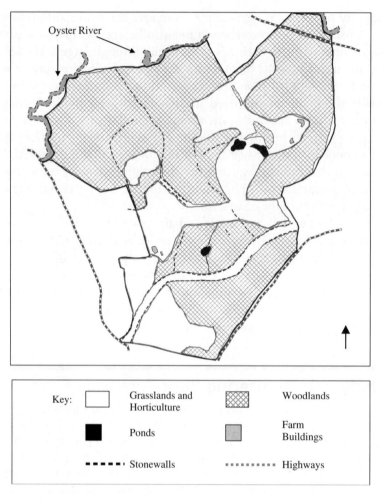

Oyster River

Key:

☐ Grasslands and Horticulture   ⊠ Woodlands

■ Ponds   ▨ Farm Buildings

▪▪▪▪▪ Stonewalls   ∙∙∙∙∙∙∙∙∙∙ Highways

Map 3.5 Distribution of woodlands and grasslands at Sheltering Rock and Tuckaway Farm, New Hampshire. Not to scale for illustration only

is turned over to haymaking, the farm's biggest cash crop and made up of orchard, Timothy and fescue grasses plus clover and alfalfa for their nitrogen fixing services. They also have 5 acres of soft fruits on rotation, and strips of sweetcorn, potatoes, sunflowers and pumpkins rotated with some of the grassland (Figure 3.19), and they keep grass paddocks for horses used for hunting. They have a number of local outlets for their crops and try where possible to recycle materials from the land especially the timber. This has been used for the farm buildings, woodchips (some of which goes to the local power plant) and woodash for mulch and fertiliser – this helps to keep up the potassium (K) and phosphorus (P) levels in the soil as well as the pH which has a tendency to be a little low for their requirements.

The grassland on the farm is maintained in a number of ways. Highland cattle are used for conservation grazing which prevents tree saplings from establishing (Figure 3.20). Timber

Figure 3.19   Grass-vegetable rotation strips. (Photo: Stephen Burchett)

Figure 3.20   Highland cattle are used to maintain open grass patches. (Photo: Stephen Burchett)

harvesting maintains patches of grass and keeps them open, which is essential for wildlife; and haymaking and crop rotations keep the land 'worked' preventing ingress of tree species.

Trees, however, are still an important component of the farm as a wildlife habitat. Many birds such as the woodcock (*Scolopax minor*) roost in trees but forage and display on open grassland. As Chuck showed us around the farm, we were taken by his impressive knowledge of the wildlife present, its requirements and his enthusiasm to ensure that the Cox farm would remain a wildlife haven for generations to come. Table 3.9 lists the wildlife that

Table 3.9   Wildlife on Sheltering Rock-Tuckaway Farm

	Common name	Latin name	Notes
Trees	Hickory	*Carya spp.*	
	Sugar Maple	*Acer saccharum*	
	Oak	*Quercus Spp.*	
	White Pine	*Pinus strobus*	
	American Basswood	*Tilia americana*	
	Dogwood	*Cornus spp.*	
Other Plants	Wood Anemone	*Anemone quinquefolia*	
	Yellow Flag Iris	*Iris pseudocorus*	
	Common Blue Violet	*Viola pedata*	
	-	*Vaccinium spp.*	
	Dense undertorey of woodland plants not identified during visit		
Invertebrates	Ringed Bog Hunter Dragonfly	*Williamsonia linterneri*	
	Karner Blue Butterfly	*Lycaeides melissa samuelis*	Endangered species
Birds	Red Tailed Hawk	*Buteo jamaicensis*	
	American Kestrel	*Falco sparverius*	
	Bobolink	*Dolichonyx oryzivorus*	
	Meadow Lark	*Sturnella magna*	
	Savanna Sparrow	*Passerculus sandwichensis*	
	Chipping Sparrow	*Spizella passerina*	
	Purple Finch	*Carpodacus purpureus*	
	American Robin	*Turdus migratorius*	
	Red Winged Blackbird	*Agelaius phoeniceus*	
	Wild Turkey	*Meleagris gallopavo*	
	Great Blue Heron	*Ardea herodias*	
	Double Crested Cormorant	*Phalacrocorax auritus*	
Reptiles	Eastern Milk Snake	*Lampropeltis triangulum triangulum*	Chuck did not specify snake species
	Ribbon Snake	*Thamnophis sauritus*	seen on the farm,
	Common Garter Snake	*Thamnophis sirtalis*	but these are likely candidates

Table 3.9   (*continued*)

	Common name	Latin name	Notes
Amphibians	Wood frog Salamander	*Rana sylvatica* -	(Figure 3.21) Eggs found in vernal pools unable to identify species
Mammals	White Tailed Deer	*Odocoilueus virginianus*	
	New England Cotton Tailed Rabbit	*Sylvilagus transitionalis*	
	Jumping Mouse	*Zapus hudsonius*	
	Coyotes	*Canis latrans*	
	American Red Squirrel	*Tamiasciurus hudsonicus*	
	Grey Squirrel	*Sciurus carolinensis*	
	Southern Flying Squirrel	*Glaucomis volans*	
	Porcupine	*Erethizon dorsatum*	Evidence of tree damage
	Fisher Weasel	*Nartes pennanti*	

Figure 3.21   Wood frog. (Photo: Sarah Burchett)

Chuck discussed during our visit although many more species would be present or visit. Birds are particularly prevalent such as savanna sparrows (*Passerculus sandwichensis*) and bobolinks (*Dolichonyx oryzivorus*) but there are also, on account of the ponds and vernal pools, wood frogs (*Rana sylvatica*) in the pools (Figure 3.21) and there was evidence of salamanders, though we were unable to identify the species from the spawn.

Lowland native grasslands of Australia

Cited in 2009 as 'Critically Endangered' and now fully protected by the Environmental Protection and Biodiversity Conservation (EPBC Act) 1999, these native lowland grasslands are one of the rarest and most endangered habitats in the world. They also support some very rare endemic species whose fates are therefore bound to them such as the striped legless lizard (*Delma impar*), the golden sun moth (*Synemon plana*), which is EPBC Listed Critically Endangered, and the rare plains wanderer (*Pedionomus torquatus*). More information can be seen about this unusual bird in Species Box 3.6. Historically covering a number of regions in Tasmania, Victoria and the southernmost parts of New South Wales, only remnants of the original habitat now exist, possibly less than 1% though more conservative estimates suggest this is more in the region of 10%. This was almost certainly natural grassland; a study in the Gippsland Plain in Victoria suggests that, although aboriginal people did burn the grassland, and that this would have certainly affected plant composition, it was unlikely to have been a principal factor in the region's treelessness. Indeed topographical, hydrological and biochemical factors are more likely to have maintained the grassy habitat, (Lunt, 1997).

Loss of this grassland habitat arose after European settlement when colonisers brought in grazing stock and exotic grassland species such as perennial ryegrass. These grasses were fertilised and watered and would have quickly out-competed the native grasses, this plus damage caused by the stock would have changed the soil structure, pH and nutrient component and local biodiversity will have been lost. Such invasions of alien species do not come as complete ecosystem packages, grasses for stock grazing are sewn as 'clean' seed, with no accompanying microbiota, weed species, invertebrates or of course the higher animals normally associated with them. So when local biodiversity is compromised as well, the habitat is effectively sterile or at best favours a few hardier local species that have been able to adapt. In the long term, this kind of system may be prone to collapse and the farmer may find that increasing costly inputs may be the only way to maintain yields – not good for the farmer and definitely not good for the local environment. The reintroduction of native species to at least some of the land, especially in marginal areas that are more difficult to farm, may begin to redress this balance. Native grasses would attract native biodiversity, an increase in biodiversity equates to an increase in biological services, for example regulation of the microclimate and local hydrological processes, recycling of nutrients, suppression of undesirable organisms and detoxification of noxious chemicals on the land (Altieri, 1999).

Species Box 3.6: Plains Wanderer (*Pedionomus torquatus*)

Profile:

Small quail-like bird endemic to eastern parts of Australia. Only member of the Pedionomidae family, and measuring between 15–19 cm, the males are brown speckled, the females are slightly larger than the males and are reddish-brown with a distinct black and white dotted collar. They are ground nesting and are very fussy about their habitat preferring tussocky grasses with bare patches. The less distinct males incubate the clutch of 4 eggs. Plains Wanderers eat seeds, leaves and invertebrates, and are in turn preyed upon by foxes.

Photo: Plains Wanderer © Geoff Jones 2009 www.barraimaging.com.au

Conservation Status:

- ICUN Red List, Endangered 2007.

- Endangered under Schedule 1, Threatened Species Conservation Act 1995.

- Critically Endangered under the Advisory List of Threatened Vertebrate Fauna (Victoria 2003).

- Vulnerable under the Environment Protection and Biodiversity Conservation Act 1999.

- Vulnerable under the Action Plan for Australian Birds 2000.

Current Status:

In general decline due to habitat loss brought about by changes in land use. Thought to be less than 8000 left. Main stronghold the Riverina of southern New South Wales.

References

www.environment.gov.au
www.birdsaustralia.com.au
http://creagus.home.montereybay.com

After significant study, a number of such restoration programmes are underway designed to aid the recovery of this ecosystem to a state as close to natural as possible.

One study used tubestock planting to try to re-establish native grassland on three separate sites. Though it was found that these plants established well, second-generation seedling recruitment was very rare. The author therefore concludes that as a result of this initial failure, conservation and management of the remnant populations of these grasses is crucial until ecologists can fully establish the requirements of the tubestock seedlings (Morgan, 1998).

An extensive project entitled 'Integrating Biodiversity Conservation into Sustainable Grazing Systems' and headed up by Professor JB Kirkpatrick of the University of Tasmania (Utas) was a four year project from 2 to 6 June whereby Prof. Kirkpatrick and his team worked with woolgrowers to look at the impact of different grazing regimes and burning on biodiversity in Tasmania. The research indicated that light to moderate grazing on unfertilised runs influenced the abundance of native species, though not necessarily the overall species composition; whereas heavily grazed areas increased bare ground patches which encouraged the ingress of opportunist annual species. In some instances even light grazing caused the exclusion of some native grassland forbs, whereas other species actually required grazing activity to survive, an interesting adaptation in an area that has only been farmed for less than 200 years! It seems that only small changes in grazing regimes result in different floral assemblages. What becomes clear is that there is no one answer to best practice management. Each landowner has, or will need, to develop a specific regime to suit the demands of their land. Overall, the report concludes that wool production and conservation in this habitat can be compatible, and notably that more than half of the woolgrowers in this region who have native vegetation on their land have implemented National Resource Management (NRM) measures within their farming practice (Leonard and Kirkpatrick, 2004; Kirkpatrick and Bridle, 2006).

Notably Professor Kirkpatrick co-authored a report of a study carried out on Department of Defence Land indicating that mowing and removal of the slash gives more positive results for biodiversity than grazing, probably as fewer bare patches were available for annual species to germinate; an interesting point and possibly worthy of implementing in some of the smaller remnants of grassland, though somewhat impractical on a larger scale, (Verrier and Kirkpatrick, 2005). It does however raise an interesting point about conservation requirements. On the one hand, as Professor Kirkpatrick points out, bare ground encourages ingress by weedy annual plants thus compromising development of healthy native perennial species. However, bare ground is a requirement for the plains wanderer who naturally lives in the grassland habitat, (see Species Box 3.1). These birds are very fussy, they require as an optimum 50% bare ground, 10% litter and 40% of the native tussocky grasses and forbs (www.environment.gov.au – accessed October 2009). Re-establishing, maintaining and protecting this vulnerable habitat is clearly a complex issue that continues to raise more questions than answers.

Eurasian steppes

Steppe grassland is found in many temperate regions of the world, though the most extensive steppes can be found in Eurasia. Vast plains stretch from the Ukraine, across Russia, Turkmenistan, Uzbekistan and Kazakhstan. Like so many natural grasslands, many regions of steppe have been ploughed up and resewn with nutrient-hungry exotic grass species. This large-scale agricultural development began in the early 1900s and heralded a significant downturn in native biodiversity. Any benefits to stock grazing were inevitably short-lived as the natural ecosystem diminished. In recent years the impetus to restore the steppes has been gathering momentum both for the benefits of regional and local biodiversity and for the long-term benefit to agricultural land management.

A number of groups are now working to this end. For example, in 1999 the Biodiversity Conservation Centre (BCC) joined with the Natural Ecosystem Laboratory (Samara) and with the Siberian Environmental Centre (Novosibirsk) in a combined programme to preserve biodiversity and promote sustainable land management through a number of projects across the former Soviet Union (www.biodiversity.ru – accessed October 2009). Their remit covers a number of ecosystem types including the steppes.

Natural steppe grassland is particularly high in forb content, and particularly in legumes and herbs such as rabbitfoot clover (*Trifolium arvense*), field brome (*Bromus arvensis*), Turkistan alyssum (*Alyssum turkestanium*) and stinking hawksbeard (*Crepis foetida*), a particularly rare species of hawksbeard. These provide a good mix of proteins, vitamins and micronutrients that promote healthy growth and development in grazing animals, (www.steppe.org.ua – accessed October 2009). Overstocking and poor husbandry in the past led to depletion in resources, however, the modern approach to sustainable land management combined with conservation initiatives should ensure that successful farming can be practised on native steppe grassland. One such initiative stretches across three countries: Russia, Ukraine and Moldova. This project, entitled 'Sustainable Integrated Land Use of Eurasian Steppe', is funded by the European Union and uses the 'Agro-Steppe Method' of planting. Natural restoration of steppe can take an estimated 80–100 years, particularly where the remaining native steppe is highly fragmented reducing opportunities for the ingress by wild species. There are two types of planting methods to help this restoration along: (i) 'surface' whereby native grassland species are inter-planted into the existing grassland and (ii) 'deep' whereby the land is reploughed and replanted with native species (www.steppe.org.ua – accessed October 2009). The geographical region is vast, therefore species diversity will vary between locations and grassland species have to be specifically selected as appropriate to the locality.

Historically not all of the steppe land that was ploughed up was resown with grasses, much was turned over to crops, but even here another approach can be used. This is to plant steppe grasslands on marginal land and as protective 'belts' around agricultural land, this increases biodiversity and still allows farmers to grow their crops, and it also helps by supplying localised biological services. A project was undertaken to form steppe shelterbelts on sloping land. These belts were approximately 7–10 m between forest belts and up to 30 m on severely eroded sand-loam slopes. The seed mix used was harvested from natural steppe and was sown without any fertilisers. Sewing was carried out in spring and autumn, though spring-sewn seed germinated and developed better than the autumn-sewn seed. These shelter-belts proved to have unlimited longevity, were self-reproducing in rich flora, regenerated annually after fire, were an optimal habitat for wildlife and significantly brought about a 95–97% reduction in run-off, (Dzybov, 2007).

The above examples scarcely scrape the surface of the works being carried out around the world to protect grassland habitats whilst at the same time accommodating the need to farm the land. What they do indicate is that conservation and farming, can and indeed does, combine to the benefit of all. Though we have a long way to go, it is hoped that some of this work will inspire those landowners who currently see environmental issues and working the land for a living as opposing forces.

References

Altieri, M.A. (1999) The ecological role of biodiversity in agroecosystems. *Agriculture Ecosystem and Environment*, **74**, 19–31.

Beldon P, (2000). Chalk Grasslands. Habitat Action Plan for Sussex.

(1998) *British Sheep*, 9th edn, The National Sheep Association.

Clyde, M.E. (2008) *Grasslands: Habitat Stewardship Series, New Hampshire Wildlife Action Plan*, The University of New Hampshire Cooperative Extension Sustainable Forestry Initiative.

Dartmoor Biodiversity Steering Group (2001) *Action for Wildlife – The Dartmoor Biodiversity Action Plan*, English Nature.

Dartmoor National Park Authority (2001) *Action for Wildlife; The Dartmoor Biodiversity Action Plan*, Dartmoor National Park Authority, Parke, Devon.

DEFRA (2000) *Fertiliser Recommendations for Agricultural and Horticultural Crops*, RB209, TSO, London.

DEFRA (2006) Hill Farm Allowance. Explanatory Booklet 2006. Available at http://www.defra.gov.uk/corporate/docs/forms/erdp/hfas/hfaguide-2006.pdf (accessed 15 November 2009).

Dzybov, D.S. (2007) Steppe-field shelterbelts: a new factor in ecological stabilization and sustainable development of agrolandscapes. *Russian Agricultural Sciences*, **33** (2), 133–135.

English Nature (2001) *The Nature of Dartmoor. A Biodiversity Profile*, 2nd edn, English Nature and Dartmoor Park Authority, Parke, Bovey Tracey.

GAP (2001) *Grazing Animals Project. The Breed Profiles Handbook: A Guide to the selection of Livestock Breeds for Grazing Wildlife Sites*. Available at http://www.grazinganimalsproject.org.uk/breed_profiles_handbook.html accessed March 2010.

GAP (2007) *Bracken Control, GAP News*, pgs 47–49. Available at http://www.grazinganimalsproject.org.uk/gap_news_archive.html accessed October 2009.

Henson, J.M. (2008) Enhancing Conversion of switchgrass to biofuels using the cellulose systems in Thermotoga neopolitana and Manatee Intestinal Tract Microbes. Ongoing project started October 2008 – Clemson University.

Haines-Young, R.H., Barr, C.J., Black, H.I. *et al.* (2000) *Accounting for Nature: Assessing Habitats in the UK Countryside*, Queens's Printer and Controller of HMSO, Norwich.

Hall, S.J.G. and Clutton-Brock, J. (1989) *Two Hundred Years of British Farm Livestock*, British Museum.

Jackson, D.L. and Jackson, L.L. (2002) *The Farm as a Natural Habitat*, Island Press, Washington, DC.

Kirkpatrick, J.B. and Bridle, K.L. (2006) *Project Report: Integrating Biodiversity Conservation into Sustainable Grazing Systems*, Land and Water Australia and Australian Wool Innovation, Ltd.

Leonard, S.W.J. and Kirkpatrick, J.B. (2004) Effects of grazing management and environmental factors on native grassland and grassy woodland, north Midlands, Tasmania. *Australian Journal of Botany*, **54** (4), 529–542.

Lunt, I. (1997) The distribution and environmental relationships of native grassland on the lowland gippsland plain, Victoria: an historical study. *Australian Geographical Studies*, **35** (2), 140–152.

McCracken Peck, R. (1990) *Land of the Eagle*, Guild Publishing, Brighton.

Morgan, J.W. (1998) Have tubestock plantings successfully established populations of rare grassland species into reintroduction sites in western Victoria. *Biological Conservation*, **89** (3), 235–243.

Price, E.A.C. (2003) *Lowland Grasslands and Heathland Habitats*, Routledge, London.

Sadava, D., Heller, H.C., Orians, G.H. *et al.* (2006) *Life: The Science of Biology*, 8th edn, Sinauers Associates Inc and Freeman and Company, Gordonsville.

Schmer, M.R., Vogel, K.P., Mitchell, R.B. and Perring, R.K. (2008) Net energy of cellulosic ethanol from switchgrass. Proceedings of the National Academy of Sciences, No 105, pp. 464–469.

Skinner, R.H. and Adler, P.R. (2009) CO_2 Sequestration Potential of Switchgrass Managed for Bioenergy Production. Extension Fact Sheets P1.

Sodhi, N.S., Brook, B.W. and Bradshaw, C.J.A. (2007) *Tropical Conservation Biology*, Blackwell Publishing Ltd, Oxford.

Trollope, W.S.W. and Trollope, L.A. (2004) Prescribed Burning in African Grasslands and Savannas for Wildlife Management. Arid Land Newsletter, No.55, May/June 2004.

UKBAP(1998), UK Crop Wild Relative Action Plans, prepared by Rosalind Codd, University of Birmingham, available at, http://www.grfa.org.uk/media_files/publications_plant/crop_wild_relative_action_plan.pdf accessed October 2009.

Verrier, F.J. and Kirkpatrick, J.B. (2005) Frequent mowing is better than grazing for the conservation value of lowland tussock grassland at Pontville, Tasmania. *Austral Ecology*, **30**, 74–78.

www.americanprairie.org (accessed 4 December 2009).

www.ars.usda.gov (accessed 21 November 2009).

www.biodiversity.ru (accessed 21 November 2009).

www.blueplanetbiome.org (accessed 4 December 2009).

www.butterfly-conservation.org (accessed 15 November 2009).

www.cotswoldseeds.com (accessed 22 November 2009

www.defenders.org (accessed 21 November 2009).

www.environment.gov.au (accessed 21 November 2009).

www.globalchange.gov, Ecoregion: Prairie Grasslands Case Study (accessed 4 December 2009).

www.lewa.org (accessed 4 Decemeber 2009).

www.mt.nrcs.usda.gov (accessed 4 December 2009).

www.nature.org (accessed 4 December 2009).

www.steppe.org.ua (accessed 21 November 2009).

www.whiteparkcattle.org (accessed 4 December 2009).

4 Forestry and conservation

4.1 Introduction

Many traditional agriculturalists would not consider forestry as agriculture and indeed in the purest sense forestry is very different to agriculture, which can be readily demonstrated by the continuous rapid conversion of the world's forest to large-scale areas of field crops and grasslands for grazing (Figure 4.1). The practice of land conversion has had a devastating impact on forest ecosystems and the wider ecological services forests provide, an in depth study of forestry will soon illustrate this point. There can be no doubt forestry is a form of agriculture, for example the global production of unprocessed wood is approximately 1.66×10^9 m^3 per year for round wood, 4.24×10^8 m^3 per year for sawnwood and 1.9×10^9 m^3 per year for wood fuel (FAO, 2009b), therefore it soon becomes very clear that forests do provide a direct consumable global crop. The continued exploitation of the world's forests must shift from one of indiscriminate exploitation and conversion to cash crops to a philosophy of sustainable forest management (SFM) because if SFM is not achieved the extent and ecology of the worlds' forests will soon collapse.

Every year the world's forests shrink by approximately 7.8 million ha (FAO, 2009a) and, although forest cover still represents 30% of the Earth's land area, at this steady rate of exploitation we could see a world with no forest cover in just over 400 years. However, simulation models of population growth and deforestation illustrate a direct correlation between human population increase and deforestation, but because of the rate of this increase the time span to total deforestation could be considerably less (Table 4.1) (Pahari and Murai, 1997). More alarmingly the tropical rainforest in Brazil could be gone in just over 50 years (National Geographic, 2001) and more widely within 75–100 years, again the cause of this devastation is directly linked to increased human populations.

Over the last 60 years continued and rapid increase in global population has occurred mainly in developing continents such as Africa, Asia and South America and consequently governments of countries in these continents are striving to raise the standard of living for their people. Countries like Brazil and Malaysia have developed at astonishing rates and are subsequently placing huge commercial pressure on the rainforests within their boundaries. These pressures originate from several sectors including subsistence and shifting agriculture to direct logging and conversion to

Introduction to Wildlife Conservation in Farming   Edited by Stephen Burchett and Sarah Burchett
© 2011 John Wiley & Sons, Ltd

Table 4.1 Population growth and deforestation model for tropical forested regions and Europe

Region	Population scenario (millions)			Forest area ('000 ha)			Population vs deforestation: R^2
	1990	2025	% Gain	1990	2025	% loss	
Tropical Asia	1559	2650	70.0	294 589	231 931	21.2	0.799
Tropical Latin America	244	382	56.6	810 831	685 416	15.5	0.813
Tropical Africa	273	683	150.0	371 006	255 255	32.0	0.847
Tropical Central America/Mexico	135	217	60.7	79 002	62 802	20.5	0.908
Sahelin Africa	1278	3260	155.0	94 663	66 622	29.6	0.799
Europe	762	762	0.0	932 355	929 839	0.3	0.718

Source: Adapted from Pahari and Murai (1997).

major cash crops such as oil palm, rubber and soya. With respect to wildlife conservation the most damaging is land conversion to cash crops, however, it is too easy to criticise such activity and lest we forget that in Europe and the developed western societies land conversion from forest cover to agriculture took place several thousand years ago.

In contrast, in the UK, forest cover has remained reasonably constant at around 10% cover for the last few hundred years, but the species composition of many UK forests has changed dramatically and native broadleaf species such as oak (*Quercus spp.*), elm (*Ulmus spp.*) and ash (*Fraxinus spp.*) have been replaced by non native species such as sycamore and plantation conifers dominated by spruce (Forestry Commission, 2003) (Figure 4.2). Again, this has resulted in significant declines in regional biodiversity and furthermore the introduction of invasive shrubby species, such as Rhododendron (*Rhododendron ponticum*) in England, Wales and Scotland and on the isle of Lundy continues to threaten the ecology of native species. Historically UK broadleaved woods were managed under a system of SFM known as coppicing and this practice has resulted in the co-evolution of native species with human exploitation, however, after the expansion of modern global economics brought about the collapse of local communities coppice practice declined and now only survives in small regional sectors and thus further impoverishes native flora and fauna of UK woodlands.

These examples illustrate how forestry is, in fact, a mimic of agriculture and in this context this chapter considers all forest (natural, logged and plantations) that are exploited by humanity as agro-forestry, and presents case studies that illustrate how SFM can increase forest cover, improve forest quality and the socio-economic status of forest communities around the world.

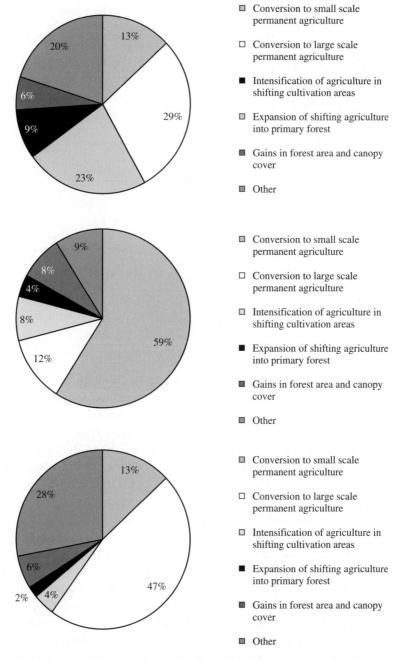

Figure 4.1 Conservation of tropical forests to other land use (a) Tropical Asia and Pacific, (b) Africa and (c) South America. (Source: FAO (2009a))

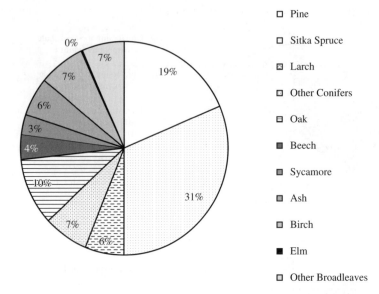

Figure 4.2 Woodland cover by principle species for UK woodlands. (Adapted from Forestry Commission (2003))

4.2   Forest management

Forest management is a complex multi-sector activity, ranging from small shifting agricultural and forest foraging by communities of indigenous forest people to modern commercial exploitation by a range of local woodland owners and major corporate companies. National government commissions such as the Forestry Commission (FC), international agencies like the International Tropical Timber Organisation (ITTO) and certification agencies like the Forest Stewardship Council (FSC) often oversee this activity. The involvement of all these sectors is crucial for the successful conservation of the world's forest and consequently forest management extends beyond the boundaries of pure ecological philosophy. The rationale for this claim is obvious; the rate of growth of the human population and their subsequent development into advanced consumer societies has direct implications for the consumption of the world's forest. Consequently the development of an interdisciplinary approach that builds on international cooperation is essential to secure a policy framework and funding for sustainable forest management. Numerous funding initiatives have evolved over the last two decades which include European LIFE projects, ITTO projects, FC funding, government tax initiatives such as the Malaysian tax relief for enrichment planting, the IKEA funded enrichment planting project in Sabah, Borneo and more recently enrichment funded by carbon credits.

These enrichment projects highlight a key concept; forests, in particular old growth forests, are important sinks for carbon. Carbon dioxide is stored in living woody tissue and in decomposing matter in the leaf litter and soil; this process is exceptionally slow in northern climates compared to tropical ecosystems. Therefore forests provide an essential ecosystem service, which can be difficult to quantify. The extent of old growth forest, in the boreal and temperate regions of the world, is 6×10^8 ha, and these forests are estimated to sequester 1.3 ± 0.5 Gtonnes of carbon per year (Luyssaert et al., 2008).

All these initiatives have had significant impact on modern forest management and consequently wider implications for conservation. However to understand how important these advances in forest management have been, a review of historic forest exploitation is essential for a wider depth of knowledge and appreciation of the fine points of modern forestry and wildlife conservation.

4.3 Forest management techniques: the UK model

The model of contemporary forest management techniques applied across the UK was closely followed by member states of Europe and the USA. In Europe and the UK, forest exploitation dates back several thousand years (Rackham, 2006). Starting just after the last glaciations some 12 000 years ago when the climate became favourable for tree growth and the vegetation of the British Isles, which had been scraped off by the ice, became replaced as seeds moved up over the land bridge from Europe before it was flooded and became the English Channel. What followed was a sequence of small-scale human settlements with Neolithic man (3800 BC) initiating settled culture with crops, livestock and timber houses. These early settlements may have exploited natural areas of grassland interspersed between the extensive tracks of the wildwood (Vera, 2000). The inevitable evolution of UK and European woodlands followed the development of human communities through time and evolved into a practice of land clearance and coppice woodland, providing a supply of useful round wood and construction timber for expanding agrarian and feudal societies. In the New World forest exploitation followed a very similar pattern but dates back to just a few hundred years as a consequence of European and colonial settlements. The importance of this historical context for wildlife conservation is illustrated below in the discussions on coppice management.

In the UK, Rackham (1988) describes three traditional management systems for broadleaf woodland:

- Woodland producing timber from uncut maidens and standards and from the under wood, mostly as coppice.
- Wood pasture where trees were pollarded or groups of trees fenced off.
- Non-woodland trees. These could be hedgerow and field trees and may have been managed using pollarding techniques or coppice to produce firewood and/or timber.

From the sixteenth century onwards plantations were introduced into the UK for timber production (Read and Frater, 1999) with an emphasis on sweet chestnut (*Castanea sativa*). Another woodland category in the UK is Royal Forests; these were managed using coppice techniques but were used by royalty for the hunting of deer and for the construction of the wooden fleet of ships such as those which participated in the fight against Armada. This last category is a direct consequence of the feudal system and perhaps represents some of the first type of designated habitats, where famous forests like Epping Forest and the New Forest still remain more or less intact today. Royal forests do not exist in the New World and in New Zealand much of the ancient forests were cleared by firing by the Maori as they hunted indigenous species such as the Moa to extinction, thus what remains is secondary forest with pockets of ancient trees.

Plantation forestry in the UK has developed into large scale extensive tracts of non-native softwoods and is managed using clear fell, selection high forest system and shelter wood system and these approaches will be discussed in context below.

4.4 Coppice

Coppice dates back to Neolithic times and evidence still exists in the UK of this early woodland exploitation, such as Grimspound on Dartmoor, which eventually became unsustainable as the climate cooled and was ultimately abandoned to moorland. Another example of early woodmanship can be observed in the track systems laid across the Somerset moors, examples include the Abbots Way (Hill-Cottingham *et al.*, 2006) which connected the islands of Burtle and Westhay and the Sweet Track (named after Ray Sweet who discovered it) which crossed 2 km of reed swamp from the island of Westhay to the dry Polden Ridge and at Flag Fen near Peterborough (Hill-Cottingham *et al.*, 2006). Pollen analysis of the surrounding peat bogs identifies the wood used in these tracks as a mixture of oak (*Quercus spp.*), lime (*Tilia spp.*), ash (*Fraxinus excelsior*) and hazel (*Corylus avellana*), which are typical coppice species. The above examples relied on a continuous production of poles that could be easily worked with primitive hand tools, and the best way to provide ample supplies of small poles is through coppicing.

Coppicing is a practice that manipulates the natural regenerative powers of a number of broadleaf species (Table 4.2), which can be cut close to the ground on a rotational cycle. The cut stumps, known as stools, will produce new shoots, often called spring, and as the stools still retain their root systems regrowth is rapid with rates as much as 3 m (10 ft) in the first season for sallow (*Salix cinerea*) and 1.5 m (5 ft) for ash (Tabor, 2000) and even oak will produce 2.1 m (7 ft) in the first season (Read and Frater, 1999). The advantage of coppicing is that it is a rotational practice and wood lots, or coupes, were cut on a 5–10 year cycle, within a landscape of a mature canopy, creating a matrix of glades and closed canopy that can be exploited by both flora and fauna.

Table 4.2 Common coppice species and end use for UK woodlands

Genus	Common name	End use
Alnus	Alder	Tool handles, bobbins, hat box, broom handles, faggots, small turnery, river piles, clogs and charcoal.
Malus	Apple	Turnery, gear teeth, mallets, firewood and charcoal.
Fraxinus	Ash	Cart wheel felloes, hop poles, masts for yachts, sports equipment, car and plane frames, barrel hoops, wattle for houses, furniture, rake heads and handles, broom and scythe handles, gates, hurdles, ladders, fencing, early weapons such as spears and arrows.
Populus	Aspen	Fruit boxes, tiles, wooden pumps, besom handles.
Betula	Birch	Plywood, faggots, handless besom broom, cotton reels, firewood, charcoal, small tool handles, tanning (bark).
Populus	Black Poplar	Limited but used for floor boards in horse drawn carts.
Ulmus	Elm	Turnery, water wheel paddles, pit wood and harbour pilings, firewood, hurdles, bean rods and pea sticks.
Corylus	Hazel	Numerous and consequently widely utilised. Thatching spars, hurdles, barrel hoops, trackway drainage, tool handles, garden furniture, faggots, numerous animal traps, wattle for wattle and daub walls, ox-yoke, hedging stakes.
Carpinus	Hornbeam	A dense hard wood and consequently used in engineering, gear teeth, roller bushes, piano keys, firewood and charcoal.
Tilia	Lime	Pea sticks, hop poles, joinery, turnery, tool handles, carving wood.
Acer	Maple	Turnery and veneers, musical instruments, kitchen utensils, platters and bread boards, chemists pestles and firewood.
Quercus	Oak	Naves for cart wheels, house and ship building, wind and water mills, tanning (bark), furniture, turnery, gates and fence post, pegs and dowels, wedges, ladder rungs and pit props.
Salix	Sallow	As hazel numerous use including cricket bats, tally sticks, wire haulage drums, brake blocks and flooring for carts.
Castanea	Sweet chestnut	Walking sticks, hop poles, fencing, ladder rungs, gate posts, window sills, pea sticks, charcoal.
Acer	Sycamore	As maple.

Source: Adapted from Read and Frater (1999) and Tabor (2000).

There are four main types of coppice producing characteristic organised woodland plots that are obviously man made, however, the composition of tree species and understorey flora is diverse, thus they represent an important anthropogenically modified habitat which has a key role to play in modern efforts to conserve wildlife.

Simple coppice: where coupes are completely cut down leaving an even age stand of stools. Species composition was traditionally mixed but as the practice of coppice became more commercialised single species stands evolved, such as sweet chestnut (*Castanea sativa*) and hazel (*Corylus avellana*).

Coppice with standards: where coupes are completely cut down using the same method outlined above, but timber trees, usually oak, are left uncut for between three and eight coppice cycles. The oak trees left standing after a coppice cut were called standards and would form a source of lumber for housing and shipbuilding. Woodsman would select mature standards with an eye to their growth form. In particular, where the primary branches joined the bole, forming natural crucks, which could be incorporated into the roof structures of buildings and the framework of the hulls in timber framed ships. These natural growth forms imparted huge strength to these early structures. The coppice species within this system tended to be hazel. The advantage of coppice with standards in the coupe was composed of a mixed age stand of trees affording niche separation for both flora and fauna.

Stored coppice: this is essentially neglected coppice and became very popular as coppicing declined. In this system one stem, the maiden, is selected and the remaining stems are removed and as the woodland matures it begins to resemble a high forest. Once the maiden has established height apical dominance mechanisms suppress regrowth of the cut side shoots. Much of the woodland in southwest England has characteristics of stored coppice.

Short rotation coppice: describes a woodland where the coppice rotation is under 10 years, typically 5–7 years, and is cut for small wood produce such as hurdle making. Species composition is usually willow (*Salix spp.*) and hazel. In modern UK forestry short rotation coppice (SRC) is undergoing a renaissance as a source of biofuel and is dominated by willow and poplar hybrids, and when increases in the area of planting for SRC is correlated with other developments in UK biofuel production, SRC will have a significant impact on the UK landscape in the future (Parliament UK, 2008).

As the market for traditional woodland products (Table 4.2) has been superceded by modern polypropylene goods (is this the era of polypropylene man) coppicing is in decline and this has implications on woodland management and biodiversity, as many organisms have adapted to coppice management practice. Many species have declined in their range since the cessation of widespread coppicing and well known examples include species such as Fritillary butterflies (*Argynnis adippe*) and violets (*Viola riviniana*). In some places coppicing has been re-introduced for the production of charcoal or to ensure the retention of a varied ground flora. In West Dean woodsite it is the population of the Wild Daffodils (**Narcissus psudonarcissus**), which along with other plants that have benefited.

4.5   Coppice and wildlife

Coppice woodlands have a very distinct life cycle based on the cutting rotation. Typically the canopy will be fully open for between one and five years and during this time the understorey vegetation will take advantage of the light which reaches the woodland floor before it gradually changes from herbaceous plants to scrub as the light diminishes and the canopy closes over (Table 4.3). From about the fifth year the canopy starts to close and the understorey vegetation is shaded out. This is due to the growth of the new shoots, as the new wood develops height and the stools spread out laterally, through extensive side branching from the stool. This period is known as canopy closure. From about the eighth year, post-coppicing, the canopy becomes completely closed and the understorey fully shaded and only shade adapted understorey plants maintain a foothold (Table 4.3). The dynamic nature of coppice canopies affords opportunities for an extensive range of different flora and fauna (Table 4.4). This is further enhanced by other habitats in coppice woodland, such as the banks, rides (trackways) and staging areas. These places are often open or in a sequential series of canopy development and form a linear network of vigorous open habitats that aid the dispersal of plants and animals. Seed is dispersed through these linear features; lepidoptera, aves and small mammals use these corridors to move from one open glade to another. These features also provide refugia for shade intolerant species like the dog violet (*Viola riviniana*), which are an important food source for the larval stage of Pearl-bordered fritillary (*Boloria euphrosyne*) butterflies.

There are three main habitat features missing in active coppice woodlands, especially in simple and SRC, these are climbers, epiphytes and deadwood. Both standing and fallen deadwood form important habitats for a number of saproxylic species, a term that describes organisms associated with deadwood, examples include fungi, fungus gnats, detritivorous insects like the springtails (order collembola) and millipedes (class diplopoda) such as *Cylindriulus punctatus*. The presence of all these saproxylic invertebrates attracts a range of carnivorous insects including the centipedes (class chilopoda) and beetles (order coleoptera). The lack of deadwood reduces habitat for saproxylic species and subsequently reduces the number of carnivorous species. In woodlands where coppicing has ceased for more than 30 years woodland and conservation managers need to make a strategic decision on whether to reintroduce coppicing or not and this will be based on the overall size and species composition of the woodland. Much of the woodland in Cornwall reflects the demise of coppicing after the First World War.

Simple and SRC also reduces the abundance and diversity of climbers and epiphytes, such as dog rose (*Rosa canina*), honeysuckle (*Lonicera periclymenum*) and even ivy (*Hedera helix*) also ferns like common polypody fern (*Polypodium vulgare*), mosses and numerous lichens. The complete harvest of all the standing wood on short cutting cycles reduces colonisation opportunity for many climbers and epiphytes and eventually exhausts their rootstocks and spores and consequently the diversity and abundance of these species decline. Climbers form an important habitat for many other species, including bats and dormice (*Muscardinus avellanarius*). For example, ivy provides important cover for spotted flycatchers (*Musciapa striata*), hawfinch (*Coccothraustes coccothraustes*) and firecrest (*Regulus ignicapillus*) and the ripening berries provide valuable food resources in early spring for thrushes (*Turdus spp.*) (Syme and Currie, 2005). Unlike many woodland plants ivy flowers in the autumn and provides a source of nectar for numerous invertebrates, especially important at that time of year, which further attracts insectivorous birds. Ivy is also valuable as a

Table 4.3 Herbaceous flora associated with coppice woodlands

Herbaceous flora

Scientific name	Common name	Preferred position in coppice woodland
Adoxa moschatellina	Moschatel	NG to EG
Allium ursinum	Ransoms	OC
Anemone nemorosa	Anemone	NG to OC
Calluna vulgaris	Heather	NG, EG, declines from MG
Campanula latifolia	Giant Bellflower	EG to MG, declines in OC
Carex pendula	Pendulous Sedge	NG to MG
Circaea lutetiana	Enchanters Night Shade	OC
Conopodium majus	Pignut	NG to EG
Convallaria majalis	Lilly of the Valley	OC
Digitalis purpurea	Foxglove	NG to EG
Epilobium angustifolium	Rosebay Willowherb	NG to EG
Filipendula ulmaria	Meadowsweet	NG to EG
Galeobdolon luteum	Yellow Archangel	NG to OC
Hyacinthoides non-scripta	Bluebell	NG to OC
Luzula sylvatica	Great Woodrush	NG to OC
Melampyrum pratense	Cow Wheat	OC
Mercurialis perennis	Dogs Mercury	OC
Narcissus pseudonarcissus	Wild Daffodil	NG to EG
Orchis mascula	Early Purple Orchid	NG, EG, but can persist in preferential sites in OC
Oxalis acetosella	Wood Sorrel	OC
Paris quadrifolia	Herb Paris	EG to MG, declines in OC
Primula elatior	Oxlip	NG to MG. Declines under persistent deep shade
Primula vulgaris	Common Primrose	NG to MG. Declines under persistent deep shade
Pteridium aquilinum	Bracken	NG to MG
Ranunculus auricomus	Goldilocks Buttercup	EG to MG, declines in OC
Rubus fruticosus	Bramble	EG to MG, declines in OC
Sanicle europaea	Sanicle	OC
Vaccinium myrtillus	Bilberry	NG, EG, declines from MG
Veronica montana	Wood Speedwell	NG to EG
Viola reichenbachiana	Early Dog Violet	NG, EG, declines from MG
Viola riviniana	Common Dog Violet	NG, EG, declines from MG

Key: NG = New growth (0–1 year), EG = Early growth (2–5 years), MG = Medium growth (5–10 years), OC = Old coppice (10–15 years) and S = Coppice with Standard.
Sources: Akeroyd (2002), Rackham (2006), Read and Frater (1999) and Rose (1981).

Table 4.4 Bird and butterfly species associated with UK coppice woodlands

Scientific name	Common name	Exploitation of coppice woodland
(a)		
Falco tinnunculus	Kestrel	NG_{fgn} and S_{fg}
Scolopax rusticola	Woodcock	NG,EG,MG_{fgn} and S_{fg}
Streptopelia turtur	Turtle Dove	EG to OC_{fgn}
Asio otus	Long-eared Owl	EG to OC_{fgn}
Caprimulgus europaeus	Nightjar	NG to EG_{fgn} and MG, OC, S_{fg}
Dendrocopos minor	Lesser Spotted Woodpecker	MG to OC_{fgn} and S_{fg}
Lullula arborea	Woodlark	NG_{fgn}
Anthus trivialis	Tree Pipit	NG to MG_{fg}
Prunella modularis	Dunnock	EG to MG_{fgn}
Luscinia megarhynchos	Nightingale	NG_{fg} and EG to MG_{fgn}
Phoenicurus phoenicurus	Redstart	NG_{fg} and S_{fgn}
Turdus philomelos	Song Thrush	NGfg, EG, MG, OC, Sfgn
Sylvia curruca	Lesser Whitethroat	EG to MG_{fgn}
Phylloscopus sibilatrix	Wood Warbler	OC_{fgn} and S_{fg}
Phylloscopus trochilus	Willow Warbler	EG and MGfgn
Muscicapa striata	Spotted Flycatcher	S_{fgn}
Parus palustris	Marsh Tit	OC_{fgn} and S_{fg}
Parus montanus	Willow Tit	MG to OC_{fgn}
Sturnus vulgaris	Starling	S_{fgn}
Passer montanus	Tree Sparrow	S_{fgn}
Pyrrhula pyrrhula	Bulfinch	MG to OC_{fgn}
Emberiza schoeniclus	Reed Bunting	NG_{fg} and EG to MG_{fgn}
Emberiza citrinella	Yellowhammer	NG_{fg} and EG to MG_{fgn}

Key: NG = New growth (0–1 year), EG = Early growth (2–5 year), MG = Medium growth
(5–10 year), OC = Old coppice (10–15 year), S = Coppice with Standard, fg = Foraging
and fgn = Foraging and Nesting.
Source: Syme and Currie (2005).

(b)				
Scientific name	Common name	Exploitation of coppice woodland	Larval food plant	Conservation status
Leptidea sinapis	Wood White	EG to MG, R	Vetch/trefoil	UKBAP
Pieris napi	Green-veined White	EG to MG	Brassica	Not listed
Anthocharis cardamines	Orange Tip	EG, R	Brassica	Butterfly conservation priority; low
Argynnis paphia	Silver Washed Fritillary	EG to MG	Viola	UKBAP

(continued overleaf)

Table 4.4 (*continued*)

(b)

Scientific name	Common name	Exploitation of coppice woodland	Larval food plant	Conservation status
Pararge aegeria	Speckled Wood	EG to MG	Poaceae	Butterfly conservation priority; low
Limenitis camilla	White Admiral	EG to MG	Honeysuckle	Butterfly conservation priority; low
Melitaea athalia	Heath Fritillary	NG and R	Cow wheat/ plantains	UKBAP
Argynnis adippe	High Brown Fritillary	NG and R	Viola	UKBAP
Boloria euphrosyne	Pearl Bordered Fritillary	NG and R	Viola	UKBAP
Satyrium pruni	Black Hairstreak	OC and S	Blackthorn	UKBAP
Thecla betulae	Brown Hairstreak	OC and S	Blackthorn	UKBAP
Apatura iris	Purple Emperor	OC and S	Sallow	UKBAP
Neozephyrus quercus	Purple Hairstreak	OC and S	Oak	Not listed
Satyrium w-album	White-letter Hairstreak	OC and S	Elm	Butterfly conservation priority; medium
Callophrys rubi	Green Hairstreak	NG, EG, R	Trefoil, gorse, heathers	Butterfly conservation priority; low
Polygonia c-album	Comma	M	Nettle/hop/ currant	Butterfly conservation priority; low
Hamearis lucina	Duke of Burgundy Fritillary	M	Cowslip	UKBAP
	Gatekeeper	M	Poaceae	–
Celastrina argiolus	Holly Blue	M	Holly/Ivy	Butterfly conservation priority; low

Key: NG = New growth (0–1 year), EG = Early growth (2–5 years), MG = Medium growth (5–10 years), OC = Old coppice (10–15 years), S = Coppice with Standard, R =Rides and M = Margins.
Sources: Read and Frater (1999) and Asher *et al.* (2002).

roost for bats (Syme and Currie, 2005). Dormice use the bark of honeysuckle to line their nests and the White admiral (*Limenitis camilla*) butterfly feeds on the flowers during June and August (Read and Frater, 1999).

The decline in woodland climbers and other epiphytic flora, in active coppiced woodland, consequently reduces habitat diversity and directly impacts a number of other fauna. Careful management of woodland rides (trackways) and the distribution and rotation of active coupes throughout the woodland can alleviate this. However there is a scarcity in fundamental research on the dynamics of both epiphytic and climber colonisation in both active coppice and other forest types, including tropical rainforest (Sutton, Borneo Books, personal communication 2008). Therefore the key to enhancement is diversity in woodland structure and landscape and accordingly woodland and conservation managers need to apply a strategic decision-making process before initiating a harvest.

4.6 Wood pasture

Wood pasture is a neglected habitat. Agronomists view wood pastures as a poor substitute for improved grassland and foresters have a similar philosophy with respect to economic viability, many pure ecologists view wood pasture as a manufactured habitat and thus not worthy of their due consideration. Consequently as a result of changing agricultural practice there has been a steady decline in woodland grazing over the last 80–100 years and therefore the habitat quality of wood pasture has deteriorated.

Put simply wood pasture is a result of grazing woodland and in England and across Europe probably originates from Neolithic livestock husbandry (Rackham, 2006). By the Anglo-Saxon period wood pasture had developed into an advanced system of wooded commons where local people had grazing rights and by about 1086 wood pastures were one of the most prominent woodlands across UK (Rackham, 2006). Wood pasture can be found throughout Europe and examples include maquis and cork oak forests in southern European countries like Spain and Portugal. These habitats were historically grazed by merino sheep and goats and have developed a range of ecological associations between the vegetation communities and regionally distinct fauna. Many pure ecologists have tried to distinguish between the European, anthropogenically-produced habitat, known as wood pasture, with savannahs found in countries like Africa and America. Academic scholars consider savannah to be natural open grassland with some trees maintained by the natural process of drought, cold and fire, the three main environmental drives for evolution of grasslands (Willis and McElwain, 2002). However, there is ample evidence to illustrate how ancient native people have used fire and slash and burn processes to manage their lands (Rackham, 2006) and consequently savannah and wood pasture are synonymous in habitat character and dynamics and consequently very difficult to apply any distinction between wholly natural vegetation and that influenced by past human practice.

Historically, wood pasture was grazed by deer, cattle, sheep, goats and ponies and in the Royal Parks deer were an important component and continue to be responsible for large scale grazing damage to woodlands today. Grazing woodlands results in a very distinct habitat, trees develop a browse line and shade tolerant and shade avoiding species are replaced by grassland, heathland (*Erica and Vaccinium spp.*), meadow type flora and some scrub. During the height of land cover by wood pasture, wood production was still very important and wood pasture could be managed by two methods, either by exclusion using compartments, a distinctly English practice or by pollarding (Rackham, 2006). Both methods will reduce grazing damage to trees and encourage natural regeneration.

Pollarding is a method where trees are coppiced above the browsing height of cattle and deer, around 2.4 m. Trees are normally pollarded from about 20 to 25 years of age and then every 12–15 years. The act of pollarding removes apical dominance and initiates the dormant lateral side shoots to sprout, and continuous cycles of pollarding result in longevity. It is not uncommon for a pollarded beech (*Fagus sylvatica*) to exceed 400 years of age. However, following the widespread decline in the practice of wood pasture there has been a subsequent decline in pollarding and many old pollards have been neglected and are now in terminal decay. The weight of the crown begins to exceed the capacity of the union joint, the joint between the crown and the bowl of the tree, and eventually the crown splits away from the bowl. This is exacerbated by fungal decay in the old heartwood of the union joint. Recognition of the biological value of old pollards has renewed interest in wood pasture by conservation biologists and in the UK conservation efforts are focused on enhancing the biological value of neglected wood pasture. In particular grazing is commonly used in conjunction with scrub cutting to control scrub encroachment. Species such as gorse (*Ulex gallii* and *U europaeus*), bracken (*Pteridium aquilinum*), young holly (*Ilex aquifolium*) and even hawthorn (*Crataegus monogyna*) will rapidly infill neglected wood pasture and subsequently out-compete some of the more beneficial grassland and heathland species. Both open grassland, with tussocks, and heathland provide a range of habitats for invertebrate species, which in turn attract insectivorous birds. Scrub encroachment will eventually develop into a dense thicket and result in an overall decline in habitat quality; consequently organisations such as the RSPB and Natural England encourage controlled grazing in old wood pasture. In Scotland the Sunart Oakwoods Initiative (SOI) is experimenting with grazing as a form of scrub management in newly opened rides and in areas cleared of commercial conifer.

4.7 Commercial forestry

The UK, like most other countries in the world uses large volumes of timber, consumption in 2005 was 44.7 million m^3 (Forestry Commission, 2006) and currently 80% of this is imported representing a significant loss of rural business activity. Since 1911, the UK has developed a substantial softwood industry based on imported exotic conifer species planted out in extensive plantations, typically in the uplands with the largest plantations occupying significant tracts of upland sites in the North of England, Wales and Scotland. The future trends in timber production from these forests are set to increase from 10.9 million m^3 during 2002–2005 to 15.5 million m^3

in the period 2017–2021 (Smith, Gilbert and Coppock, 2000). This large volume of timber requires a high degree of commercial management and consequently has both negative and positive implications for wildlife conservation. Below we discuss briefly how commercial forestry is managed and consider some of the implications for biodiversity.

Commercial forestry is based on plantations, which are essentially large tracts of land given over to the production of timber, normally from imported quick growing species, in the UK this is dominated by Sitka spruce (*Picea sitchensis*). In tropical countries like Malaysia commercial plantations have been established, initially in the states of Kedah and Selangor in the late 1950s, based on African mahogany (*Khaya ivorensis*) (Yahya *et al.*, 2006). These exotic, hardwood plantations are also managed on an intense commercial basis and, like their UK counterparts, have significant implications for wildlife. Consequently understanding how plantations are managed will enable conservation managers to target their efforts on achievable objectives and to work with plantation managers in a sustained effort to improve the overall landscape scale conservation of biodiversity.

4.8   High forest

Commercial forestry has been designed to maximise timber yield and quality with minimal effort and, as such, one silvicultural system dominates the industry, which is the High Forest system. High Forest occurs where a stand of trees is grown from seedlings that are all planted out at the same time, and eventually all harvested at once. This leads to an even age stand of trees; generally conifers but can include broadleaved trees. There are numerous environmental disadvantages with this system. When a stand is felled it leads to a rapid and extreme change in the microclimate which gives rise to soil erosion, desiccation and weedy plant colonisation by opportunistic species such as Rhododendron (*R. ponticum*) (Read and Frater, 1999; Hart, 1991). Furthermore, unlike coppice, the stand size tends to be very large and since many modern high forests were planted alongside lochs and other inland water bodies, soil erosion will, where it occurs, be extensive and may lead to negative impacts on the water quality of adjacent water bodies.

4.9   Planting

Planting, establishment and early forest maintenance is a complex science and a full description is beyond the scope of this text (Hart, 1991). However, for the sake of a comprehensive understanding of the production cycle, of a coniferous high forest, an example of forest establishment is discussed below for Sitka spruce.

Sitka spruce is the most commonly planted commercial conifer in the UK and is marketed as a low-density construction grade timber. In such cases the load bearing capacity of Sitka spruce declines as planting distances increase and it is not recommended that Sitka spruce be planted beyond 2 m × 2 m spacing. Thus the number of trees per hectare will equate to 2500 seedlings in a square planting design (Hart, 1991) and where a triangle planting design is used 15.5% more plants are required (Hart, 1991). Other conifers such as larch (*Larix decidua*) and Douglas fir (*Pseudotsuga menziesii*) are known as high-density timber species and can be planted out at a wider spacing (Hart, 1991). The planting season is usually from October to early April in the lowlands and from October to early June in the uplands, especially Scotland (Hart, 1991). Planting depth is important and varies according to the site. In upland peat and gleys the water table is within 10 cm for most of the year and thus seedlings must be planted on a ridge or next to an old stump or root plate and not too deep (Hart, 1991). Where the soil is free draining then the seedlings can be notched into the planting site and on droughty soils plants must be planted 5 cm below the root collar. Applications of fertiliser, in particular Phosphate (P), can aid early growth and 7 5 kg of P/ha have been applied to heath soils in southern England and in Wales, in deep upland peat, 50 kg P/ha has been used. In general though, fertiliser is not used in upland conifer afforestation and restocking with the exception of Nitrogen (N) in heather checked spruce (Hart, 1991).

The growth and establishment of young plantations can be checked by excessive weed growth, in particular grasses, broadleaved weeds such as Rosebay Willowherb (*Epilobium. angustifolium*) bramble (*Rubus fruticosus*), gorse, broom (*Genisteae spp.*), bracken, heather and woody regrowth all compete for space, light, nutrients and water. Control is mainly through pre-planting cultivations and/or chemical control. In the UK all pesticide operations must follow the Control of Substances Hazardous to Health Regulations (CoSHH) 1988.

The most challenging weeds to control are the grasses and bracken, because ploughing gives poor control and both plant groups are vigorous competitors. Grasses are generally controlled by herbicides but bracken often requires a dual approach where an application of herbicide is used (Glyphosate or Asulam). When bracken is controlled by herbicide application in the presence of a tree crop a follow up of hand cutting is required one month later. This is to prevent the tall fronds collapsing over the young tree seedlings.

4.10 Harvesting

There are several approaches to harvesting timber and, in economic terms, the most favoured is clear fell. This technique leads to an even age stand of trees and large gaps that are often exploited by species such as the woodlark (*Lullula arborea*), nightingales (*Luscinia megarhynchos*) and Nightjar (*Caprimulgus europaeus*) (Hart, (1991); Read and Frater, (1999); Syme and Currie, 2005). In contrast to clear fell, the irregular system provides continuous cover and is preferred by conservationists (Hart, 1991).

Figure 4.3 Township of An t-Aoineadh Mor on land owned by Forestry Commission for Scotland, Inniemore. This site was uncovered by tree felling in 1994. (Photo: Sarah Burchett)

This system leads to less soil erosion and can provide a range of age classes and habitat and improve the aesthetic value of the landscape. The Irregular system relies on uneven-age selection of trees for harvesting and consequently an uneven-age of forest regeneration. It is typically more time consuming to manage on a commercial basis and in the USA would be a more appropriate system for non-industrial forest owners (NIFOs).

Another relatively benign logging method is the selection forest system. Here individual trees (single selection) or small groups of trees, up to 0.1 ha (group selection) are felled. Natural regeneration or deliberate planting then fills the gaps. Single selection relies on shade tolerant species such as beech (*Fagus sylvatica*) and some conifers and thus is not widely adopted, but is a useful approach for logging operations in sensitive areas, such as historic sites that were over-planted by commercial crops (Figure 4.3) and by sensitive watersheds.

Lastly the shelterwood system is an approach where groups or strips of trees are felled across a stand. The remaining trees then act as a shelter for replanting and this system is an appropriate practice in exposed upland sites. The shelterwood system can be beneficial to wildlife as there are always a number of trees left standing, and even plantation conifers provide habitats for a number of species, including the crossbill, (*Loxia curvirostra*) firecrest, (*Regulus ignicapillus*) goldcrest (*R. regulus*) and black redstart (*Phoenicurus ochruros*) (Syme and Currie, 2005).

4.11 Sustainable forest management

The basic principle of SFM is that forests produce their products and services in perpetuity with no undue loss of their inherent values (ITTO, 2007). It should be possible to apply SFM to all of the worlds' forests including temperate and tropical forests, where the basic principle is that if trees are felled and their timber extracted, new trees are planted, which over time perform the same functions as the original forest. Globally this does not always occur and damaging forest management still occurs across substantial tracts of forest landscapes. The main reason for this is that SFM is very difficult to define and assess. In order to address this problem the ITTO have developed a clear definition of SFM that can be applied across the world's forests.

SFM is *'the process of managing permanent forest land to achieve one or more clearly specified objectives of management with regard to production of a continuous flow of desired forest products and services without undue reduction in its inherent values and future productivity and without undue undesirable effect on the physical and social environment'* (ITTO, 2007).

In keeping with the definition a brief overview of SFM is as follows and in the context of conservation biology priority is given to environmental enhancement.

1. SFM must maintain ecosystem services including the provision of clean water, soil structure, biology and stability, conservation of biological organisms and the sequestration of carbon.
2. SFM must extend to and maintain the economic sustainability of the global forestry sector, be transparent and improve the distribution and trade of legal timber.
3. SFM must be financially viable and its introduction must improve the social and economic status of communities dependent on the forest industry.

In order to monitor the implementation of above model the ITTO have developed a set of seven criteria and associated indicators (ITTO, 2009).

The ITTO apply the above elements to trade in tropical timber but globally these criteria can be used to improve sustainable forestry in temperate forest and in context of the case studies presented here SFM has many similarities in the desired vision and outcomes.

In this chapter we present four case studies that illustrate how applications of SFM philosophy can improve both the socio-economic status of the world's forestry and, at the same time, maintain and enhance the biological value of forests and consequently wildlife conservation. The following four case studies represent large scale initiatives supported by national and international policy, advisory bodies, research and funding.

1. **The Sunart Oakwoods Initiative (SOI).** This is a large scale European and agency-funded project in the west of Scotland that aims to improve the

conservation value of Atlantic oak woods by improving the socio-economic status of marginal rural communities.

2. **Long leaf pine (LLP)** (*Pinus palustris*). LLP was once a widespread native forest type extending across the southeastern states of America, which were felled during the expansion of white America. Replanting, where it occurred, was with another faster growing timber species known as Loblolly pine (*Pinus taeda*) and consequently the extent and quality of LLP forests has declined to a few remaining pockets. The United States Department of Agriculture (USDA) extension service oversees the conservation of these remaining pockets and this case study illustrates how successful economic markets for LLP needles is a key element to the future survival of this forest type and when carried out in partnership with advisors from USDA extension service will enhance forest cover.

3. **Broad leaf forest of New Hampshire (NH).** NH is a state on the northeastern seaboard of America where the landmass is currently covered by 80% forest. This has implications for migratory birds as complete forest cover removes important habitat features such as glades, rides and grassland and the challenge in this forest sector is biodiversity enhancement and maintaining an economically viable forestry industry. Extension officers from the USDA and the University of New Hampshire work with local landowners to support projects that improve both biological diversity and financial viability. The case study explores how forestry is integrated into the wider farming practice and demonstrates how the application of traditional coppice management in conjunction with beaver dams improves diversity in the landscape matrix.

4. **Malaysian enrichment planting programme.** Extensive and rapid logging and land conversion of the tropical rainforest in South East Asia has had a huge detrimental impact on species distribution and survival, not the least some very important commercial tropical timber trees. The Malaysian government, in partnership with numerous interested parties, has instigated a number of forest enrichment planting projects that have the potential to restore degraded permanent forest and enhance biological diversity. The case study presented here illustrates some very important advances in enrichment planting.

Case Study 4.1: Sunart Oakwoods Initiative

On the isolated Ardnamurchan peninsular in the far western region of the Scottish Highlands there is a unique community-based initiative to restore Atlantic oakwoods. Atlantic oakwoods were highlighted as a European habitat of conservation concern, following the 1992 EU Habitats Directive, and in 1995 a EU LIFE sponsored project initiated the origins of the SOI, which has now expanded to the Morvern and Moidart regions (Map 4.1). Similar community woodland projects can be found on the Little Assynt Estate, such as the Culag Community Wood.

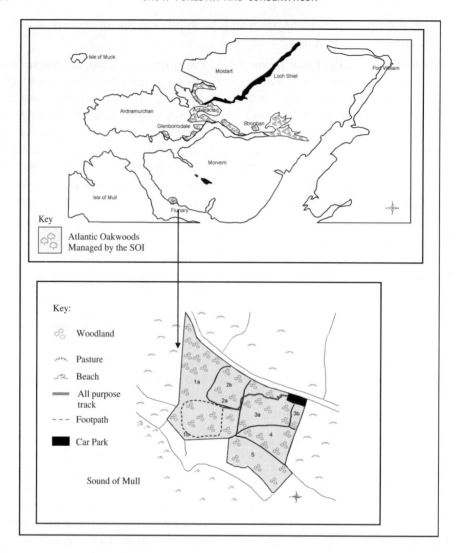

Map 4.1   Location and extent of the Atlantic Oakwoods managed under the Sunart Oakwood Initiative and detailed map of the Achnaha Community Woodland. Map not to scale and for illustration only

Atlantic Oakwoods

Atlantic oakwoods are restricted to the Atlantic coastal fringes of Europe occurring in countries such as Britain, France and Spain. The prevailing climate is one typical of western regions in that it is damp and humid. Atlantic oakwoods are described in the UK Biodiversity Action Plan as upland oakwoods and are also referred to as Britain's temperate rainforest. This concept is easy to appreciate when one walks through this type of woodland because the tree boles, limbs, stumps, decaying material and the remnant walls and dykes are festooned with numerous species of bryophytes, club mosses and ferns (Table 4.5) many of

Table 4.5 A selection of lower plants associated with SOI

Level of taxa	Scientific name	Common name	Notes
Club Moss	*Lycopodium annotinum*	Interrupted Clubmoss	Habitats Directive: Annex 2/4/5 and NS
	Lycopodium clavatum	Stag's-horn Clubmoss	Habitats Directive: Annex 2/4/5 and EC,CITES schedule D
Liverworts	*Acrobolbus wilsonii*	Wilson's Pouchwort	NS, BAP 2007
	Adelanthus decipiens	Deceptive Featherwort	NS
	Anastrophyllum hellerianum	Heller's Notchwort	NS
	Calypogeia integristipula	Meylan's Pouchwort	NS
	Calypogeia suecica	Swedish Pouchwort	NS
	Cephalozia catenulata	Chain Pincerwort	NS
	Cladopodiella francisci	Holt Notchwort	NS
	Cryptothallus mirabilis	Ghostwort	NS
	Eremonotus myriocarpus	Clubwort	NS
	Fossombronia foveolata	Pitted Frillwort	NS, BAP 2007
	Haplomitrium hookeri	Hooker's Flapwort	NS
	Jamesoniella autumnalis	Autumn Flapwort	NS
	Jungermannia borealis	Northern Flapwort	NS
	Jungermannia confertissima	Kidney Flapwort	NS
	Jungermannia subelliptica	Two-lipped Flapwort	NS
	Lophozia longidens	Horned Flapwort	NS
	Lophozia obtusa	Obtuse Notchwort	NS
	Nardia geoscyphus	Earth-cup Flapwort	NS
	Plagiochila atlantica	Western Featherwort	NS
	Plagiochila carringtonii	Carrington's Featherwort	NS
	Radula voluta	Pale Scalewort	NS
	Riccardia incurvata	Lesser Germanderwort	NS
	Scapania calcicola	Calcicolous Earwort	NS
	Sphenolobopsis pearsonii	Horsehair Threadwort	NS
	Tritomaria exsecta	Cut Notchwort	NS
Mosses	*Bryum intermedium*	Many-seasoned Thread-moss	Red List, NS
	Bryum pallens	Pale Thread-moss	NS
	Distichium inclinatum	Inclined Distichium	NS
	Ditrichum flexicaule sensu lato		NS
	Glyphomitrium daviesii	Black-tufted Moss	NS
	Grimmia decipiens	Great Grimmia	NS
	Grimmia funalis	String Grimmia	NS
	Grimmia longirostris	North Grimmia	NS
	Hedwigia ciliata sensu lato		Red List, NS
	Trichostomum hibernicum	Irish Crisp-moss	NS
Ferns	*Hymenophyllum wilsonii*	Wilson's Filmy-fern	Red List

Source: NBN Gateway (2009).

which characterise these Atlantic oakwoods and are gazetted as rare. In the UK the area occupied by Atlantic oakwoods extends to 70 000 ha which makes them the largest area of Atlantic oakwoods in Europe. The dominant species is sessile oak (*Quercus petraea*) with secondary species composing of downy birch (*Betula pubescens*), alder (*Alnus glutinosa*), ash (*Fraxinus excelsior*), wych elm (*Ulmus glabra*), rowan (*Sorbus aucuparia*), holly (*Ilex aquifolium*), willow (*Salix spp.*) hazel (*Corylus avellana*) and bird cherry (*Prunus padus*) and occasionally pedunculate oak (*Quercus robur*). Many other species are associated with UK Atlantic oakwoods (Table 4.6) and a full National Vegetation Classification (NVC) description of the floristic community can be found in Rodwell (1998). Some of these species are very rare with many being listed as red data book (RDB) species and or nationally scarce (NS). The species list for the Sunart and Morvern oakwoods extends to over 6300 species (JNCC, 2009) but includes species that migrate into and out of the adjacent landscape matrix which includes lochs, ancient farmland sites (crofts), neighbouring mountain and open moorland sites and adjacent conifer plantations.

Table 4.6 Species associated with Atlantic oakwoods and commonly encountered in the Ardnamurchan region

Flora		Invertebrates	
Common name	Scientific name	Common name	Scientific name
Alternate Water-milfoil	*Myriophyllum alterniflorum*	Grey Scalloped Bar	*Dyscia fagaria*
Bird's-nest Orchid	*Neottia nidus-avis*	Poplar Hawk-moth	*Laothoe populi*
Bluebell	*Hyacinthoides non-scripta*	Narrow-bordered Bee Hawk-moth	*Hemaris tityus*
Blue-eyed-grass	*Sisyrinchium bermudiana*	Elephant Hawk-moth	*Deilephila elpenor*
Bog Orchid	*Hammarbya paludosa*	Buff-tip	*Phalera bucephala*
Bogbean	*Menyanthes trifoliata*	Sallow Kitten	*Furcula furcula*
Broad-leaved Helleborine	*Epipactis helleborine*	Swallow Prominent	*Pheosia tremula*
Daffodil	*Narcissus pseudonarcissus subsp. pseudonarcissus*	Coxcomb Prominent	*Ptilodon capucina*
Early-purple Orchid	*Orchis mascula*	Small Chocolate-tip	*Clostera pigra*
Field Pansy	*Viola arvensis*	Dark Tussock	*Dicallomera fascelina*
Green-winged Orchid	*Orchis morio*	Dew Footman	*Setina irrorella*
Heath Dog-violet	*Viola canina*	Beetle (Coleoptere)	*Deronectes latus*
Irish Lady's-tresses	*Spiranthes romanzoffiana*	Beetle (Coleoptere)	*Dytiscus lapponicus*
Lesser Butterfly-orchid	*Platanthera bifolia*	Beetle (Coleoptere)	*Gyrinus minutus*

Table 4.6   *(continued)*

Flora		Invertebrates	
Common name	Scientific name	Common name	Scientific name
Lesser Spearwort	*Ranunculus flammula*	Beetle (Coleoptere)	*Gyrinus opacus*
Lords-and-Ladies	*Arum maculatum*	Beetle (Coleoptere)	*Hydroporus erythrocephalus*
Marsh Cinquefoil	*Potentilla palustris*	Beetle (Coleoptere)	*Hydroporus melanarius*
Marsh Pennywort	*Hydrocotyle vulgaris*	Painted Lady	*Vanessa cardui*
Marsh Violet	*Viola palustris*	Small Tortoiseshell	*Aglais urticae*
Marsh-bedstraw	*Galium palustre*	Green Hairstreak	*Callophrys rubi*
Marsh-marigold	*Caltha palustris*	Chequered Skipper	*Carterocephalus palaemon*
Narrow-leaved Helleborine	*Cephalanthera longifolia*	Clouded Yellow	*Colias croceus*
Primrose	*Primula vulgaris*	Peacock	*Inachis io*
Purple Iris	*Iris versicolor*	Small White	*Pieris rapae*
Shoreweed	*Littorella uniflora*	Red Admiral	*Vanessa atalanta*
Small Cow Wheat	*Melampyrum sylvaticum*	Dark Green Fritillary	*Argynnis aglaja*
Snowdrop	*Galanthus nivalis*	Small Pearl-bordered Fritillary	*Boloria selene*
Star Sedge	*Carex echinata*	Marsh Fritillary	*Euphydryas aurinia*
Sweet Violet	*Viola odorata*	Small Copper	*Lycaena phlaeas*
Water Lobelia	*Lobelia dortmanna*	Green-veined White	*Pieris napi*
Yellow pimpernel	*Lysimachia nemorum*	Common Blue	*Polyommatus icarus*
(Liverworts) Wilson's Pouchwort	*Acrobolbus wilsonii*	Common Flat-backed Millipede	*Polydesmus angustus*
Fitzgerald's Notchwort	*Leiocolea fitzgeraldiae*	Millipede (Invasive species)	*Brachydesmus superus*
Atlantic Pouncewort	*Lejeunea mandonii*	Snake Millipede	*Proteroiulus fuscus*
Western Featherwort	*Plagiochila atlantica*	Striped Millipede	*Ommatoiulus sabulosus*
Carrington's Scalewort	*Radula carringtonii*	Blunt-tailed Snake Millipede	*Cylindroiulus punctatus*
Pale Scalewort	*Radula voluta*	Milipede (Invasive species)	*Cylindroiulus latestriatus*
Horsehair Threadwort	*Sphenolobopsis pearsonii*	White-legged Snake Millipede	*Tachypodoiulus niger*

Source: NBN Gateway (2009) and JNCC Taxon List and Personal Observation.

Threats to the Sunart Atlantic Oakwoods

Nationally the major threat is fragmentation due to under-planting and conversion to commercial conifer plantations and/or clearance for upland grazing. Another key threat is from alien vegetation such as Rhododendron (*Rhododendron ponticum*), introduced in the Highland region as an ornamental plant by the land-owning classes, following the 1745 uprising. The other major floristic invader is the Japanese knotweed (*Polygonum cuspidatum*) and both of these species are aggressive invaders and difficult to eradicate. The SOI has developed a very successful strategy for controlling Rhododendron, which involves uprooting the whole plant and subsequently mulching the brash with a follow up bud eradication action in the following year. This process is known as the Lever and Mulch (LaM) method and is described below.

As with many other UK woodlands, the Sunart Atlantic oakwoods have a long history of commercial exploitation based around coppicing and woodland grazing. Local crofters carried out both of these sustainable practices up until the 1950s when the FC started to establish large-scale commercial softwood plantations. This and other modern commercial pressures led to a decline in the local crofting community and thus the local commercial value of the indigenous woodland. This culminated in neglect and abandonment of traditional woodland management, which is another key threat to long-term security of these important oakwoods.

Sunart Oakwoods Initiative

The SOI covers 10 000 ha of forest, including commercial conifer plantation. The initial Sunart oakwoods project started at Strontian on the shores of Loch Sunart, subsequent expansion has resulted in the SOI extending southwards to Morvern and the sound of Mull, north to Moidart and west to the Ardnamurchan point (Map 4.1). In most places the oakwoods follow the edge of lochs and occupy sides of glens forming an important interface between land and the water's edge.

These Atlantic oakwoods have numerous conservation designations which include 14 SSSIs that cover 19 862.5 ha, three National Nature Reserves (NNRs) a special protection area (SPA) for bird life, it is a RAMSAR site and contains two National Scenic Areas (NSAs), one RSPB Nature Reserve and one Scottish Natural Heritage Nature Reserve. This list is impressive and alludes to the local community's sense of place and thus strengthens the community philosophy that underpins the restoration work being carried out.

The SOI has been developed around three key themes:

1. To establish a 25 year regeneration programme with the aim of restoring and regenerating the existing Atlantic oakwoods.

2. To develop a sustainable management strategy that would benefit the local community through employment and other enterprise activities.

3. Provide access to the oakwoods for both the local community and visitors with the aim of developing a wider understanding and knowledge of the regional biodiversity and historical context of the Sunart oakwoods.

These aspirations are to be met by targeting key areas of activity:

1. Community consultation
2. Deer control and management
3. The removal of non-native trees from the oakwoods
4. The upgrading of car parks and walks
5. The provision of woodland interpretation leaflets
6. The preparation of an oakwoods management plan
7. Recruitment of staff
8. Deliver targeted training to reduce local skills gap and to encourage development of local employment.

Woodland Management

Due to the extensive size of this project, the woodlands are being managed as a partnership between the Forestry Commission for Scotland (FCS), local landowners (estates such as Glenborrodale) and community groups such as the Morvern Community Woodland group. This approach is an essential element in securing a sustainable local economy. One major project undertaken by the North Sunart Woodland Group is the identification of three contiguous blocks of woodland to be fenced off. The aim is to reduce deer grazing on regenerating oak seedlings and the replanted native broadleaf trees. This has been achieved and to date 1300 ha of woodland on the shores of Loch Sunart have been fenced off and the non-native trees removed.

In the Sunart region a total of 28 027 m of deer fencing has been erected, 158 ha of conifers have been removed, 42 ha of rhododendrons have been cleared, 2030 m of existing woodland walks have been upgraded and 4500 m of new woodland walks created, along with access for the disabled, the construction of three new car parks and an observation hide. The impact of this work is very visual (Figure 4.4) and will in the long-term lead to the regeneration of native Atlantic oakwoods dominated by sessile oak. However, there are numerous obstacles to overcome before this vision is fully achieved. Following the removal of non-native trees and rhododendrons other invasive flora will quickly establish, including rosebay willowherb (*Epilobium angustifolium*) ragwort (*Senecio jacobaea*) and bracken and natural regeneration of Sitka spruce seedlings, which need rouging (manual pulling of weeds) This vegetation requires constant management, consisting of cutting and bashing which is labour intensive and, critically, not very sustainable once agency funding comes to an end. An alternative approach being explored by the SOI is to use highland cattle in selective grazing of the newly cleared areas. This has implications for the oak seedlings so cattle will need to be excluded from some areas to ensure the seedlings are not taken. This may seem counter productive but in reality could be more beneficial and economical than controlling the invasive flora by hand. Cattle will graze the invasive vegetation and bruise the rhizomes of bracken reducing the vigour and subsequent regrowth. Furthermore, cattle may give rise to some localised poaching (soil disturbance) and these sites form important ephemeral habitats for the aquatic larval stages of many invertebrates. Finally, in keeping with the community philosophy and sustaining local economic growth, these cattle could be marketed at local restaurants, tourist centres and accommodation providers.

Figure 4.4   Visual impact of conifer clearing within the SOI project area. Note remnant oak trees in the background and thick stand of young birch in the foreground. (Photo: Stephen Burchett)

Another approach being applied by the SOI is once the sitka spruce has been removed, natural regeneration is dominated by birch (*Betula spp.*) and currently the method is to allow thick stands of birch to develop (Figure 4.5) which help to protect oak seedlings from deer predation. This method will also help to suppress the invasive vegetation but does not significantly impact on the health of oak seedlings. The current thinking is that the birch will complete its pioneer life cycle and eventually the slower longer-lived oak will form the climax community. On our visit we saw several healthy oak seedlings in the understorey of the regenerating birch but have to note that many of the remnant, mature oaks, that were incorporated in to the commercial conifer plantation were in poor health. Their canopies were atypical and very thinly covered in leaves. Boles were distorted and limbs growing at uncharacteristic angles. In defence of the SOI this is to be expected, as these remnant trees would have experienced high competition for recourses from the commercial softwoods. Therefore it is of paramount importance to protect and encourage strong healthy growth of the oak seedlings.

Figure 4.5 Dense stand of naturally regenerating birch, which help to reduce deer grazing on regenerating oak seedlings. (Photo: Stephen Burchett)

Rhododendron Control

Rhododendron is a non-native shrub introduced originally to UK gardens from the Himalayan areas of Nepal and India and, in milder climes, is a vigorous coloniser. Propagating via prolific seed production and stem layering, it can colonise substantial areas of land and subsequently out-compete with local flora resulting in a reduction in regional biodiversity. The standard practice for controlling rhododendron is repeated applications of herbicide and or manual removal of standing vegetation with the use of chainsaws and manual cutting. This procedure leaves the stump behind which has considerable powers of regeneration and within just a few years rhododendron can re-establish a substantial stand of above ground vegetation.

In the SOI the FCS have been working with local parishioners to develop a more sustainable approach known as the lever and mulch method. This approach results in the removal of the stump. The process starts by what is known as breaking-in, which involves removing a section of branches, stems and leaf litter so that an operator can access the centre of the plant. Once inside the shrub, the operator inspects the crown to assess the most appropriate method of removing the crown complete with root ball or reducing the crown to below root-collar. This often results in most of the above ground material being cut away and piled up in close proximity to the shrub being removed (mulching) leaving a substantial stem known as the lever. This needs to be about 1.4–1.6 m high so that operators can exert pressure. Then the stem and root ball is physically removed or reduced to below the root-collar level. Once removed the root ball is piled onto the mulch stack and turned upside down to aid desiccation of the crown material.

Mulching in close proximity to the original shrub reduces regeneration from seed and desiccating the root ball and stump material will significantly reduce vegetative regrowth.

The following spring, operators need to return to the original site and inspect the stumps, large branches and root balls for new viable buds, which, if observed are removed via a standard claw hammer. At the same time an assessment is made of seedling regeneration and any new seedlings are manually rogued out and added to the mulch stack. Follow up action is not desperately time consuming, In one 0.04 ha area with 24 shrubs, a full FCS time and motion analysis was carried out during the follow up action. The average time to treat one plant was approximately 50 seconds and average time to access the next plant was 15 seconds and the whole process for 24 shrubs in the sample area (0.04 ha) took just 22 minutes which equates to just over 9 hours for 1 ha with a potential plant population of 600 shrubs per ha.

On our visit to the same study area, just one year later, it was clearly obvious how successful the LaM method is. We observed no regrowth of rhododendron and the mulch stacks were well covered with bryophytes (Figure 4.6) and numerous areas were beginning to recover their native vegetation with a good flush of native bluebells (*Hyacinthoides non-scriptus*).

Figure 4.6 Mulch stack of *R.ponticum* following application of leaver and mulch eradication method. Photograph was taken 18 months after original application of leaver and mulch. (Photo: Sarah Burchett)

Achnaha Community Woodland

Achnaha Community Woodland is located in the Parish of Morvern some 3.5 km WNW of the village of Lochaline, overlooking the Sound of Mull (Map 4.1). The woodland extends to some 8 ha which comprises of 6.5 ha of woodland and 1.5 ha of open ground adjacent to the woodland. In the past sheep from neighbouring estates have grazed this open ground but in recent years grazing has occurred by the local deer population. This decline in planned grazing could have a detrimental impact of important flora such as orchids, as reduced grazing pressure will give rise to bracken encroachment which will subsequently out compete less competitive flora.

The underlying bedrock is composed of volcanic lavas with intruded basalt dykes, which have undergone several glacial erosions. The origins of these lavas date back to the Tertiary period when the whole region was a centre of extensive volcanic activity (Achnaha Community Woodland Management Plan, 2007) and the same geology is also seen on the Isles of Mull, Staffa, Skye and the Small Isles as well as the Giants' Causeway in Northern Ireland (Achnaha Community Woodland Management Plan, 2007). Drainage of the woodland is complicated by several factors. Firstly the very fact that Achnaha woodland is located in the predominantly maritime west coast of Scotland ensures the woodland receives high levels of precipitation. Secondly there is little variation in the underlying bedrock and the absence of any surface watercourses means the main influence on drainage is in variations in local topography. The highest point is 36 m above sea level, in the northeastern area of the wood, falling to the high tide mark on the southerly shoreline. Within this predominately southerly slopping landscape the woodland has several other local topographic variations that together govern the local vegetation. It is this local variation in topography that determines the depth, wetness, fertility and acidity of the soil. On the steep slopes shallow and well-drained soils occur. On the rocky outcrops and rocky plateaux recently formed thin soils, that are less leached, occur and on the shallow slopes deeper brown earth soils arise. In the flatter wetter areas humus rich patches occur. The soil chemistry is generally neutral to mildly acidic due to leaching of bases as a consequence of Morvern's high rainfall.

History

The historical status of the woodland is not fully described with the only firm date of woodland management being the recent felling of the spruce in 2005. There is evidence of crofting style management with stonewalls and dyke systems and speculations are presented (Achnaha Community Woodland Management Plan, 2007) that the site was traditionally a woodland croft.

Woodland Type

Almost all the mature trees have been planted and protected from deer grazing by a network of dykes and stonewalls. These trees are mostly sycamore and ash but some oak is present. Attempts to classify the woodland using the NVC have been made and local practitioners describe several different communities and associated sub-communities ranging from W11 through to a small area of W25a (Table 4.7). This large variation in woodland description illustrates the changes in local topography and soil conditions. Classification has been blurred due to the large number of non-native trees and therefore relies on the abundance of native understorey vegetation (Table 4.7).

Table 4.7 NVC communities reported for in the Achnaha community woodland

Community description	NVC code	Notes on understorey flora
Quercus pretraea-Betula pubescens-Oxalis acetosella	W11	Grasses: *Agrostis capillaris, Anthoxanthum odoratum, Poa trivialis.* Ferns: *Dryopteris dilatata, D. filix-mas, Athyrium filix-femina* and *Pteridium aquilinum.* Mossess: *Thuidium tamariscinum* and *Eurhynchium praelongum.* Herbs: *Veronica chamaedrys, Viola riviniana, Ajuga reptans* and *Veronica serpyllifolia.*
Dryopteris dilatata sub-community	W11a	Ferns more abundant growing with *Rubus fruticosus* and *R. idaeus.*
Blechnum spicant sub-community	W11b	*Primula vulgaris*
Alnus glutinosa-Fraxinu excelsior-Lysimachia nemorum	W7	Trees: *Alnus glutinosa* and *Fraxinus excelsior.* Grass: *Deschampsia cespitosa,* Herbs: *Lysimachia nemorum, Filipendula ulmaris.*
Urtica dioica sub-community	W7a	Grass: *Phalaris arundinacea.* Fern: *Chrysosplenium oppositifolium.* Mossess: *Thuidium tamariscinum* and *Eurhynchium praelongum.* Liverworts: *Lophocolea bidentata* and *Pellia epiphylla.* Herbs: *Urtica dioica.*
Carex remota-Cirsium palustre sub-community	W7b	*Equisetum arvense.* Sedge: *Carex flacca.* Mossess: *Calliergonella cuspidata, Brachythecium rivulare* and *Rhizomnium punctatum.* Liverwort: *Trichocolea tomentella.*
Deschampsia cepitosa sub-community	W7c	As W7 but with Grassess: *Dactylis glomerata, Holcus lanatus.* Rushes: *Juncus effusus.* Ferns: *Athyrium filix-femina, Dryopteris dilatata.* Moss: *Thuidium tamariscinum.* Herbs: *Ranunculus repens* and *Oxalis acetosella.*
Pteridium aquilinum-Rubus fruticosus agg underscrub, *Hyacinthoides non-scripta* sub-community	W25a	Ferns: *Pteridium aquilinum, Dryopteris filix-mas.* Grasses: *Dactylis glomerata, Holcus lanatus.* Herbs: *Urtica dioica, Heracleum sphondylium, Chamerion angustifolium, Ranunculus acris, Veronica chanaedrys, Viola riviniana* and *Hyacinthoides non-scripta.* Mossess: *Thuidium tamariscinum, Rhytidiadelphus squarrosus* and *Eurhynchium praelongum.* Liverwort: *Lophocolea bidentata.*

Source: Achnaha Community Woodland Management Plan and Personal Data.

The combination of abandonment and variations in local topography give rise to six key features:

- Veteran sycamore and ash trees with some oak
- Lichens including some notable Atlantic woodland species
- Stands of regeneration
- Craggy bank in the southwest corner
- Wet woodland in the southeast corner
- Historical features such as dykes and field boundaries.

These features give rise to local heterogeneity and thus help to enhance local biodiversity, the future management of the woodland needs to consider management options that maintain and enhance these six key characteristics.

Management Aims and Objectives

The long-term vision for the future management of Achnaha Woodland is based on three key themes:

- Balance the productive and wildlife aspects of the woodland
- Be an exemplar model of broadleaf woodland management in an area where broadleaf woodland has fallen into neglect
- Develop community benefits in terms of new skills, woodland walks, raw material and education.

To achieve these aims the steering group for Achnaha Woodland has developed two areas of operational objectives (Wildlife Conservation and Silviculture) that rely on the development of community participation from volunteering through to key skills training and local rural business supported by the FCS.

Wildlife Conservation Objectives

- No net loss of native species as a result of operations and less frequent or uncommon species should be maintained
- Trees that will never be felled should be identified, marked and mapped. Examples include elm, holly, veteran trees with high habitat value; ivy clad, natural pollards, trees with high epiphyte colonies and rare saprophytes
- Native species not present in the wood should be reintroduced where appropriate. Examples would include aspen (*Populus tremula*), yew (*Taxus baccata*), guelder rose (*Viburnum opulus*), bird cherry (*Prunus padus*)
- Encourage native shrub layer following clearance of *Rhododendron ponticum*
- Creation of a pond

- Encourage structural diversity in the woodland
- Maintain veteran trees
- Allow beech to decay and do not replace
- Plant up bank in compartment 5 (Map 4.1)
- Habitat augmentation using local public participation. Examples include establishment of nest boxes, bat boxes and rotten wood piles.

These conservation aims must align with the Silviculture and the following series of objectives have been established:

- FCS to continue low levels of deer control
- Coppice individual alder
- Thinning and coppice of sycamore and ash
- Veteran replacements by thinning for structural form
- Collaborate with FCS to deal with dangerous trees
- Weed out beech regeneration and monitor sycamore regeneration
- Leave some spruce to fully develop
- New planting in compartment 3b (Map 4.1)
- Protect individual orchard trees in car park
- Apply selective forestry techniques on semi-mature and mature trees in compartment 4. Some to be harvested for local use some to be identified for future potential (Achnaha Community Woodland Management Plan, 2007).

Woodland Management: Thinning and Coppice

Coppicing has begun in the western corner of coupe 1a (Map 4.1), with both sycamore and ash being removed for local green woodworking and charcoal burning. Regeneration of the native understorey is taking place and removal of the canopy is allowing greater light penetration to the woodland floor stimulating birch and oak seedling regeneration. Unwanted wood has been piled up to increase resources for important saproxylic species and other detritivorous organisms. The adjacent dense stand of sycamore (*Acer pseudoplatanus*) and ash is targeted for thinning in order to meet the following commercial objectives:

1. Improved stand quality by removing poorly formed (Figure 4.7), defective, damaged and diseased trees
2. Ensure future increase in the growth of the girth is concentrated in the best formed trees
3. Produce stands of revenue generating trees
4. Ensure satisfactory development of mixed stands of broad leaf trees by the timely removal of undesirable species
5. Prepare coupes for natural regeneration.

Figure 4.7 Removing defective trees such as the one illustrated in this figure improves the growth characteristics of young trees and encourages straight linear growth of the bole to the crown with minimal branching, improving the long term economic viability of the regenerating young trees. The tree illustrated above has several dead branches, is leaning into new growth and is covered in honeysuckle. Removing this tree will damage the honeysuckle but if care is taken the roots will survive and the vine will establish on a neighbouring tree. (Photo: Stephen Burchett)

Thinning practice is aimed at achieving a final stand density that ensures high quality timber trees and for the two commercial tree species in Achnaha Wood the following final crop stocking rates for sycamore are 140–170 stems/ha with a mean spacing of 8.1 m and for ash 120–150 stems/ha and a mean spacing of 8.7 m. To achieve this trees have to be thinned little and often (every 5–10 years) in the first 40–45 years and once the maximum mean growth increment has passed thinning will occur on a 10–15 year cycle.

Thinning is combined with pruning and in the first instance the first 3 m of the main stem should be pruned before the first thinning occurs and then the stem should be pruned to 5–6 m before the second thinning. The aim of pruning is to produce a tree where the crown is one third of the height of the tree. This ensures even growth and maximum commercial lumber.

During thinning, large trees that present a risk to the commercial crop because they are either poorly developed, damaged and/or occupying a large proportion of the canopy must be removed.

When carrying out thinning attention needs to be paid to the biological needs of both ash and sycamore. Ash is a strong light-demanding tree and in order for high quality timber to develop, where there are 4–16 growth rings per 25 mm, their crowns must be free from competition. This is best achieved by thinning crowns frequently to maintain a light open crown over one third of the tree. Once ash is constrained, generally by a small canopy, response to future thinning and pruning is poor. Another aspect that requires attention to detail is the control of ash canker and any trees presenting symptoms must be removed and the diseased material disposed of. Sycamore does not require this level of attention to detail as it can recover very quickly from poor management such as neglect and untimely thinning. The best time to manage sycamore is to thin heavily from an early age and during the months of February and March. To produce good quality timber trees using the above method will take between 60 and 75 years.

Understorey and Woodland Glade Management

Invasive rhododendron (*Rhododendron ponticum*) has been completely removed using the aforementioned LaM method, this has opened up the understorey, and the native flora is rapidly recovering, with evidence of bluebells (*Hyacinthoides non-scripta*), small cow wheat (*Melampyrum sylvaticum*), violets (which are important food plants for the larval stage of many fritillary butterflies) and numerous species of lower plants. However, there are a few problems with other weedy species such as willow herb, bracken and bramble that needs constant, time consuming management. Again questions arise about the long-term sustainability of manual weed control. There have been several approaches adopted which include manual weed control with aid of local volunteers, this has been very successful but a future model is one where low density woodland grazing may be used in conjunction with manual intervention, the appropriate breed being the highland cattle. If this model were to be employed cattle would only have a limited window in which to graze woodland coupes and thus would need to be finished in adjacent fields. This could be very beneficial to the adjacent landscape matrix.

Management of Local Landscape Matrix

The landscape immediately adjacent to Achnaha community woodland is neglected pasture that includes 1.5 ha of pasture that borders the western boundary of the woodland (Map 4.1). Currently the condition of the grassland sward in this pasture is reasonable due to some *ad hoc* grazing and thus still contains numerous grassland species including orchids. However the pasture systems above the woodland have been abandoned since the mid 1980s and the vegetation is dominated by bracken (Figure 4.8), which is an aggressive invasive species of grassland and in particular neglected upland pasture. Bracken has a significant detrimental impact on grassland floristic communities and once established will bring about a major decline in the diversity of herbaceous flora (Chapter 3). Another consequence of bracken, if grazing was to be resumed in these pastures, would be the considerable welfare issues to be addressed, for example the growing tips of bracken are poisonous to livestock, dense stands of bracken bring about an increase in the tick populations which transmit disease (Chapter 3) and finally bracken is not very palatable and the herbage quality is poor, consequently livestock performance will be weak, resulting in financial loss to the crofter.

Figure 4.8   Bracken encroachment in a meadow adjacent to Achnaha community woodland. Grazing was withdrawn from this meadow in the mid 1980s. (Photo: Stephen Burchett)

If recovered these neglected pastures could play an important role in enhancing local biodiversity, as they will approximate woodland glades dispersed throughout a predominantly forested landscape. If grazed at low stocking density the paddocks will develop a sward composition that is composed of a patchwork of tightly grazed swards with a mix of perennial herbaceous species and lightly grazed swards that develop into tussocks. Both these sward types act as a resource for fauna. In the matrix of tightly grazed grass there will be numerous herbaceous species that act as food plants for lepidoptera and hymenoptera and in the longer grasses and tussocks other flora like the hawkbits, (*Leontodon spp.*) and knapweeds (*Centaurea spp.*) will extend their flower heads, which will attract lepidoptera and hymenoptera during the summer months. The tussocks will also act as refugia for small mammals and other invertebrates such as orthoptera, which in turn will attract birds. Therefore it is important that the woodland does not exist in isolation and thus the management team have aspirations to tackle the restoration of these pastures in the future, but the first action will have to be an attempt to bring the bracken under control, which will require a significant input of manual labour to cut and crush the bracken fronds twice a year for several years to reduce its regenerative vigour.

Commercial Utilisation and Biodiversity

One example of how sycamore, a non-native tree, can aid diversity is that the bark of sycamore appears to be suitable for epiphytic lichens, such as the Lobarion or lungwort lichens and large *Lobarion pulmonariae* communities are commonly seen. These lichens are very sensitive to atmospheric pollution and are quite rare in the UK. These lichens require moderate light levels so have responded well to the eradication of *R. ponticum* and the light coppicing practice and subsequent understorey management, in particular around older and lichen rich trees. Lobarion lichens also colonise oak, ash, rowan and hazel but are poor colonisers of the more acidic bark of commercial conifers and rhododendron. Achnaha Community Woodlands are also rich in other lichen and bryophyte species.

On-Site Milling and Woodland Products

In order to support and achieve the above management practices these woodlands have to be set in a limited commercial context, in particular the long-term low input approach and ultimately the low volumes of high-grade timber produced would not support external contractors. Thus the underlying approach has been the development of community involvement, which starts with training and partnership support. As a consequence across the whole SOI project there has been a productive partnership developed by community woodland groups and chiefly the FCS. One positive relationship is that the FCS purchased new low impact timber extraction machinery that is capable of removing 5 tonnes of lumber with minimal damage. The machinery is on permanent loan to the woodland communities across the SOI. Another positive step is the development of on-site milling of timber with mobile bandsaw and mill. Subsequent use of the lumber in boardwalks and community projects such as the Sgilean na Coille, which is Gaelic for outdoor learning facility, in Achnaha Community Wood and further afield a Sgoil na Coille (forest school) (Sunart Oakwoods Initiative, 2009). The outdoors learning facilities enable forest managers to hold onsite training days and invite local schools groups for woodland activities such as botanical identification and fungal forays. Other positive impacts are the development of local woodturning crafts, charcoal production and the development of wood fuel power and heating.

Community Approach

It is clear that the conservation or the sustainable exploitation of the woodland cannot occur without community involvement. Labour is required for the planned coppicing and thinning programmes and a local end market is required for the woodland products, fire wood and lumber, and the planned coppicing and thinning is essential in maintaining and enhancing local biodiversity (Figure 4.9). Therefore the ultimate aim of the wider

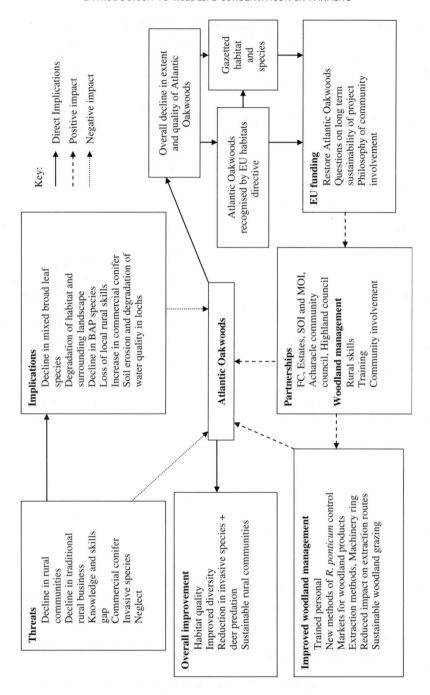

Figure 4.9 Chart of the Sunart Atlantic oakwood Initiative

Table 4.8 Political and community structure of Sunart oakwoods initiative

Partners	Funding bodies	Training	Benefits
Forestry Enterprise	LEADER	Ardnamurchan training and skills project	Improvement in pool of local rural skills
Acharacle Community Council	EU LIFE Project Funding	Chain saw	Improved local biodiversity
Crofters Commission	EU Structural Fund	Chemical spraying	Machinery ring
Forestry Commission Scotland Estates	Forestry Commission Scotland	Fencing	Nature interpretation centre
Highland Council	Highland Council	Machinery operation	New footpaths
Local Archaeology Group	Lochaber Enterprise	Wood turning	New method to control Rhododendron
Local communities: Glenborrodale, Salen, Strontian, Acharacle	Millennium Forest for Scotland Trust	–	New rural business; general arboriculture, fire wood, mobile saw Mill, Woodland crafts, wood turning studio
Local land owners	Rural Challenge Funding	–	Outdoor learning facilities
Lochaber Enterprise	–	–	Supply of local wood fuel and timber
Scottish Natural Heritage	–	–	Wider participation; school visits, disabled access, nature hide

Source: Sunart Oakwoods Initiative (2009).

SOI is the development of local skilled labour and rural business and wider community involvement through education and extension programmes. Overall the project has achieved several milestones and has demonstrated how integrated community involvement can be beneficial to woodland, wildlife conservation and the wider community (Table 4.8).

NB A particular favourite with the local community is the native red squirrel (*scuirus vulgaris*), Britain's only native squirrel. This appealing creature has a stronghold here (Species Box 4.1).

Species Box 4.1: Red Squirrel (*Sciurus vulgaris*)

Profile:

The Red Squirrel is native to the British Isles and its range also covers large areas of Europe and Northern Asia. It is a small squirrel measuring between 180 and 240 mm in length, plus a tail of a further 180 mm. Spending most of their time in the tree canopy, they eat mainly pine seeds but also nuts, fruit, shoots and fungi, often storing fungi in trees to feed on over winter as well as buried nuts. They, in turn, are preyed upon by birds of prey, pine martens and domestic cats. They build nests called 'dreys' and will have a number of these throughout their range moving from one to the next possibly to reduce parasite build up. They live for around six years,

Image: Red Squirrel © Sandra Hughes 2009

and are mature in their first year, mating in late winter to produce two to three kittens in the spring, though they sometimes produce a second litter later in the year. The kittens are weaned after 8–12 weeks when their teeth are fully developed. Although they are sometimes found in broadleaf woodland their preferred habitat is coniferous forest.

Conservation Status:

- Biodiversity Action Plan (BAP) Species
- IUCN Red List – Least Concern
- Appendix III Bern Convention

Current Status:

The animal is under threat in the UK since the introduction of the American Grey Squirrel, which out-competes the red squirrel but, they are further disadvantaged in that their population density is only 1 per ha, as opposed to the grey which has a density of 8 per ha. Their numbers have dwindled to 140 000, 85% of which live in Scotland. Most of the remaining 15% live in Cumbria, NW England, though they have a small stronghold on Brownsea Island off the southern English coast with between 200 and 250 squirrels, and a few other remaining hotspots in England and Wales. Other threats come from disease and from habitat fragmentation. Some work to limit the spread of grey squirrels in Scotland has led to increases in the population of the red squirrel.

References

www.forestry.gov.uk
www.nationaltrust.gov.uk
BAP plan

Case Study 4.2: Longleaf Pine

This case study does not focus on a specific farm but on a species of native American pine, the LLP (*Pinus palustris*) sometimes called the yellow pine. LLP forests are integral to many farms in the southern states and thus form part of the farm income, they also occur on private non-industrial forestlands and in public forest. The conservation of LLP is a key federal target and as such on-farm conservation is supported through the Continuous Conservation Reserve Programme (CCRP36) and thus one small case study will be presented where LLP is being managed for numerous objectives including the production of leaf litter mulch and as a wider habitat to support on-farm game shoots.

Distribution and Forest Structure

LLP is a distinctive forest type of the southern forest region of the United States. The southern forest region is a vast expanse of mixed forest that occurs across 13 states in the Southeastern United States. Before European settlement the southern forests covered approximately 90 million acres (36.4 million ha) but it is now very fragmented through land conversion activities such as agriculture, commercial pine plantations and urban development (Brockway *et al.*, 2005). Further degradation of the southern forest, including and very specifically LLP is caused by a political move to suppress forest fires. This is a

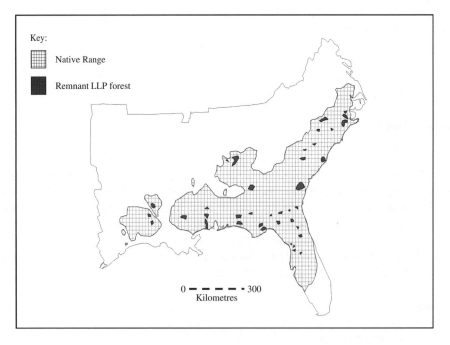

Map 4.2   Current distribution of long leaf pine superimposed onto native range distribution. (Source: Brockway *et al.* (2005) and The Longleaf Alliance (2009))

contentious issue and, where forests abut urban development, known as the Wildland-Urban interface forest management by prescribed burns is very difficult (Vince et al., 2005).

The original extent of LLP forests was around 38 million acres (15.4 million ha) (Map 4.2) stretching from southern Virginia down to mid Florida, then westwards as far as eastern Texas. Now only around 3 million acres (1.2 million ha) exist in fragmented pockets (Map 4.2). Due to the tree's tolerance of a diverse range of soil types and environmental conditions, LLP forests occupy an extensive range of landscapes (Map 4.2).

LLP forests have been broadly classified into four categories, each with their own distinctive character (Table 4.9). However, the range of plant and animals that exploit

Table 4.9 Habitat distribution and characteristics of longleaf pine forest in Eastern USA

Habitat	Elevation	Soils	Forest character	Cause of decline
Montane	Up to 600 m a.s.l, on southern and south westerly slopes	Well drained flinty gravels, sandstone ridges, rocky outcrops (quartzite, sandstone, mica, schists and gneiss)	Subtle grade from pure LLP canopy to mixed southern pines.	Fire suppression, urban development and forest conversion
Sandhill	On sandy ridges up to 181 m a.s.l.	Droughty deep white sands, low fertility and varying depth, generally from 0.6 to 1.5 m but can be up to 45 m in the Carolinas	Widely spaced parkland like appearance. Fire stunted deciduous trees and shrubs including oak. Understorey bunchgrass and herbaceous flora.	Fire suppression is major factor for stand degradation. Land conversion to sand pine
Rolling hill	From 40 to 76 m a.s.l. in rolling hill country.	Well drained fertile soils. Brown sandy loam to 375 mm depth. Fossiliferous material and some limestone parent material and outcrops	LLP canopy with mixed hardwood understorey. Bunchgrass and herbaceous.	Conversion to agriculture, urban development, fire suppression leading to invasive plants and conversion to loblolly pine
Flatwoods and Savannah	Tide line to 40 m a.s.l.	Low fertility and moderate to poor drainage. Acidic and low in organic matter. Ash grey silt clay appearance	High density of LLP, where present. High density of herbaceous plants. Swampy patches and wet prairies.	Exotic plants, fire suppression, urban development. Conversion to loblolly and slash pine

Source: The Longleaf Alliance (2009).

the native range of LLP is extensive with some species being quite tolerant of a range of habitats, for example the oaks and grass species, and other species being quite specific such as the sandhill lupine (*Lupinus perennis*) and the southern magnolia (*Magnolia grandiflora*) limited to the warmer flatwoods and savannah habitats (The Longleaf Alliance, 2009). The range of soil types where LLP forest can be found is quite staggering, in the mountain regions soils are derived from granite, quartzite, schist, slate, sandstone, limestone and dolomite parent material. These soils are generally stony and well drained. The lowland forest stands occupy soils derived from marine sediments that form deep, coarse, well-drained sands to poorly drained clays (Brockway *et al.*, 2005). As the forests expand into the rich fertile rolling hills and piedmont the soils are dominated by red-yellow ultisols. Sandwiched between the rolling hills of the lower coastal plain and the mountain range is a sandy ridge, commonly called the sandhills, where the soils are characterised by entisols. These soils are deep sands with weakly developed horizons and are characteristic of the xeric sandhill LLP forest type, that typify the stands in the southern region of North Carolina. As the LLP forest cover continues towards the coastal regions, the soils become dominated by the spodosols which are wet sandy soils where the water table is very close to the soil surface during the rainy season (Brockway *et al.*, 2005) (Map 4.3). Clearly this diverse range of soil types will determine the associated plant communities and as such give rise to a large diversity in plant species and associated animal life. Indeed LLP forests are considered by many to be the second most plant-species rich communities outside the tropics (Peet and Allard, 1993). Plant surveys have recorded as many as 140 species of plants in 1000 m^2 sample areas and 40+ species in square metre plots (Peet and Allard, 1993).

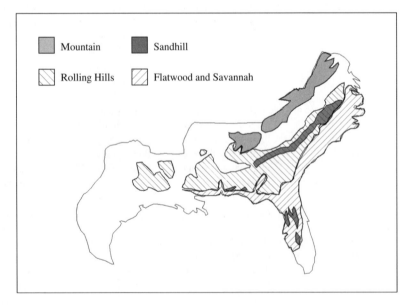

Map 4.3 Long leaf pine habitats, map not to scale for illustration only. (Source: The Longleaf Alliance (2009))

The general structure of a LLP is one of an open park like forest where uneven age mosaics of even aged patches of trees predominate. This strange mosaic of forest structure is the product of natural regeneration and lightning induced fires. The understorey of LLP is composed of coarse grasses that form partially open herbaceous understorey plant communities. In the western gulf coastal plain LLP understorey grass is dominated by bluestem grasses (*Schizachyrium scoparium* and *Andropogon spp.*) and from Florida moving north and east the understorey grasses are wiregrass (*Aristida stricta*) also known as pineland threeawn and beyrich threeawn (*A. beyrichiana*), collectively these grasses are called bunchgrasses. When LLP is in its native state it is the understorey grasses and the fallen pine needles that facilitate the ignition and spread of fire during the growing season and it is this very natural process that has shaped the natural structure of LLP forest over countless millennia. Since the spread of European settlements and the development of modern America, LLP forests have been heavily logged and, following logging, converted into modern pine plantations using Loblolly pine (*Pinus taeda*), sand pine (*Pinus clausa*) and slash pine (*Pinus elliotti*) and or converted to agriculture and more recently further degraded by extensive urban development (Brockway *et al.*, 2005).

Natural Ecosystem Process

In LLP forest the presence of dry bunchgrass and fallen pine needles leads to an understorey fuel supply technically termed a fine-fuel matrix (Brockway *et al.*, 2005). Fire can spread quickly through this mixture but seldom reaches the canopy and as the fuel is dry it burns quickly and generally at a low temperature. These low temperature burns prevent lethal damage to the root zone and cambium of the mature trees. Forest fires are common in the US and for LLP natural disturbance is the driver for natural forest regeneration. It is the variable nature of lightning strikes and tornados, arising from tropical storms, which cause many trees to fall and thus create natural gaps. LLP seedlings are shade-intolerant and thus young trees achieve favourable establishment in the centre of gaps (Brockway *et al.*, 2005). Both the abundance and growth of LLP seedlings is negatively related to mature LLP trees and a closed canopy. The mature trees exert a competitive exclusion zone on seedlings that extends to approximately 15 m into forest gaps and on xeric sands this is known as the seedling exclusion zone (SEZ) (Boyer, 1963). Following the establishment of new seedlings in gaps intraspecific competition for resources and recurring forest fires, seedling density starts to decline in the SEZ until the final natural stand size is achieved. In more mesic sites this phenomenon has not been observed and is presumably the result of ample resources characteristic of mesic sites, resulting in a higher density of trees.

   LLP seedlings are very susceptible to fire in their first year, but after this establishment year, they have evolved a strategy to cope with forest fires. Seedlings have a grass stage (Figure 4.10) during which root growth is favoured over shoot growth and consequently the seedling shoots remain as a tuft of needles surrounding a large terminal bud. This growth form results in no stem and thus no cambium is exposed to fire. The grass-like needles can be burnt off, often without killing the terminal bud, and thus seedling survival and regeneration is highly probable following a quick low temperature burn. Once the seedlings have developed sufficient root reserves, the growth form changes from a grass stage where seedlings rapidly bolt to about 2 m in one or two seasons. This change from grass growth to

Figure 4.10 Longleaf pine seedling. Kevin Williams from the USDA extension service is illustrating how fire will consume the grass like needles leaving the important bud undamaged. (Photo: Stephen Burchett)

Figure 4.11 Fire adapted bark of longleaf pine. (Photo: Stephen Burchett)

woody extension places the terminal bud out of reach of lethal fires in natural LLP habitats. Mature LLP trees have a thick bark (Figure 4.11 and Species Box 4.2) that protects the vulnerable cambium. Fire also results in natural pruning of LLP creating a clear bole between the ground and the canopy.

Species Box 4.2: Long Leaf Pine (*Pinus palustris*)

Profile:

The LLP is a medium to large tree of around 30 m in height. As the name implies, it has the longest needle shaped leaves of any pine, and these vary from 20–45 cm, as can seen in the image of the sapling. The cones are also large, measuring between 15 and 25 cm. The scaly bark is a thick red/brown. The tree is found in fairly open stands, the resulting light penetration allowing for a diverse under-storey of grasses, small shrubs and flowering plants, though there is notably very little diversity in the mid-storey.

Photo: Long Leaf Pine Sapling © Sarah Burchett 2008

Conservation Status:

- IUCN Red List – Vulnerable Status

Current Status:

Many replanting projects underway (see text)

References

www.ufl.edu

www.longleafalliance.org

However, due to fire suppression policies (Vince *et al.*, 2005) many LLP forests have developed a thick understorey of weedy hardwood species (Figure 4.12), such as oak, that fuel forest fires and which can lead to very damaging crown fires and subsequent degradation of natural ecosystem process.

Threats to Longleaf Pine Forest

LLP forest is recognised as being the most endangered of the southern ecosystems and the threats are numerous, ranging from non-sustainable commercial logging, where the logged-over forests are replanted with quick growing commercial pines like the Loblolly pine. Conversion to large-scale pine plantations is another major threat, where the area occupied by such plantations is predicted to increase by 67% to 22 million ha by the middle of the twenty-first century. Conversion to agricultural use, including small grain cereals, fruit and grasslands for cattle during the period between 1790 and 1850 occurred on the more fertile sites. The most recent threats to LLP now stem from urban development and southern states such as Florida and the Carolinas are rapidly being transformed into urban super-corridors, not too dissimilar to the multimedia corridor of Putrajaya and Cyberjaya

Figure 4.12   The open parkland characteristics of longleaf pine leads to hardwood infill (oak), which if left unmanaged, develops into a tall thicket. Note charred bark on the pine trees from an earlier controlled burn. (Photo: Stephen Burchett)

on the outskirts of Kuala Lumpur, Malaysia. Recent examples of conversion of LLP forest to other land use is overwhelming, from 1987 to 1995 37 000 ha of LLP in Florida was converted to other uses and during this eight-year period 3000 ha per year was converted to urban use and 1500 ha per year were given over to agriculture (Outcalt, 1998).

These significant urban developments have profound implications for all the forest systems in the US and are, perhaps, the most significant in terms of forest system dynamics for LLP is the suppression of fires in and around urban centres. This is of course quite understandable on the grounds of safety and protection of properties but the ecological implications for LLP forest are quite profound.

Conservation and Restoration of Longleaf Pine Forests

The large diversity in plant communities found in LLP forest have, of course, led to a diverse range of animal species, many of which are listed as endangered (Table 4.10) and in total there are 30 plant and animal species associated with LLP which are considered as threatened or endangered (Lander *et al.*, 1995). One very characteristic species associated with LLP forest, the red-cockaded woodpecker (*Picoides borealis*) is listed as endangered and only thrives in LLP which is over 60 years old (Species Box 4.3).

It is the combination of the social, economic and biological diversity, coupled with an impressive list of threatened and valuable species and the very fact the LLP still occurs across most of its natural range, although in isolated fragments (Map 4.2), that has underpinned contemporary conservation efforts.

Table 4.10 A selection of species commonly associated in Sandhill long leaf pine

(a) Sandhill-Plants-Birds

Common flora		Aves	
Scientific name	Common name	Scientific name	Common name
Amsonia ciliata	Fringed bluestar	*Colinus virginianus*	Bobwhite quail
Aristida stricta	Wiregrass	*Aimophila aestivalis*	Bachman's sparrow
Asimina incana	Pawpaw	*Buteo jamaicensis*	Red-tailed hawk
Baptisia lecontei	Pineland wild indigo	*Cathartes aura*	Turkey vulture
Callisia rosea	Sandhill roseling	*Certhia americana*	Brown creeper
Carya pallida	Sand hickory	*Columbina passerina*	Ground dove
Carya tomentosa	Mockernut hickory	*Corvidae cristata*	Bluejay
Clitoria ternatae	Butterfly pea	*Corvus brachyrhynchos*	Common crow
Diospyros.spp	Persimmon	*Dendroica pinus*	Pine warbler
Galega officinalis	Goat's rue	*Lanius ludovicianus*	Loggerhead shrike
Helianthemum corymbosa	Pinebarren frostweed	*Megascops. spp*	Screech owl
Heterotheca.spp	Golden aster	*Myiarchus crinitus*	Great crested flycatcher
Indigofera caroliniana	Carolina indigo	*Pipilo erythrophthalmus*	Rufous-sided towhee
Iris verna	Dwarf iris	*Piranga rubra*	Summer tanager
Licania michauxii	Gopher apple	*Picoides borealis*	Red-cockaded woodpecker
Lupinus perennis	Sandhill lupine	*Sialia sialis*	Eastern bluebird
Muhlenbergia capillaris	Hairawn muhly	*Sitta carolinensis*	White-breasted nuthatch
Opuntia humifusa	Pricky pear cactus	*Sitta pusilla*	Brown-headed nuthatch
Phlox nivallis	Pineland phlox	*Falco sparverius*	Southeastern kestrel
Platanthera ciliaris	Orange-fringed orchids	*Zenaida macroura*	Mourning dove
Q myrtifolia	Myrtle oak	–	–
Q. margarettae	Sand post oak	*Sciurus niger shermani*	Sherman's fox squirrel
Q. arkansana	Arkansas oak	Mammals	
Q. incana	Blackjack oak	Scientific name	Common name
Q. margarettae	Sand post oak	*Sciurus niger shermani*	Sherman's fox squirrel
Quercus falcata	Southern red oak	*Geomyidae* (several genera)	Pocket gopher

(*continued overleaf*)

Table 4.10   (*continued*)

(a) Sandhill-Plants-Birds

Common flora		Aves	
Scientific name	Common name	Scientific name	Common name
Sassafras.spp	Sassafras	*Glaucomys volans*	Southern flying squirrel
Schizachyrium.spp and *Andropogon.Spp*	Bluestem bunchgrasses	*Lasiurus cinereus*	Hoary bat
Sporobolus junceus	Piney woods dropseed	*Sigmodon hispidus*	Hispid cotton rat
Vaccinium angustifolium	Low-bush blueberry	*Sylvilagus floridanus*	Eastern cottontail rabbit
Vaccinium arboreum	Sparkleberry	*Urocyon cinereoargenteus*	Grey fox
Viola pedata	Bird's foot violet	*Vulpes vulpes*	Red fox

Adapted from Long Leaf Pine Alliance.

(b) Reptiles and amphibians

Scientific name	Common name
Gopherus.spp	Gopher tortoise
Cnemidophorus sexlineatus	Six-lined racerunner
Crotalus adamanteus	Eastern diamondback rattlesnake
E.e. lividus	Bluetail mole skink
E.e. onocrepis	Peninsula mole skink
Eumecer egregius simillis	Nothern mole skink
Eumeces egregius	Northern mole skink
Heterdon platirhinos	Eastern hognose snake
Heterdon simus	Southern hognose snake
Hyla femoralis	Pinewood's treefrog
Hyla gratiosa	Barking treefrog (or bell frog)
Masticophis flagellum	Eastern coachwhip
Ophisaurus attenuatus	Slender eastern glass lizard
Pituophis melanoleucus	Northern pine snake
Rana areolata aesopus	Florida gopher frog
Rana capito	Carolina gopher frog
Rhineura floridana	Florida worm lizard
Sceloporus undulatus	Fence lizard
Sistrurus miliarius	Pygmy rattlesnake
Tantilla relicta	Central Florida crowned snake
Tantilla relicta relicta	Peninsula crowned snake

Species Box 4.3: Red Cockaded Woodpecker (*Picoides boralis*)

Profile:

A rare, medium sized, non-migratory woodpecker found almost exclusively associated with the LLP Tree (*Pinus palustris*). The Red Cockaded Woodpecker weighs around 44 g, is 21–22 cm long and has a wingspan of 34–37 cm. It eats mainly insects and spiders plus some berries. The males excavate cavities in the live tree (unusual for woodpeckers which normally prefer dead wood). Around the cavities, small pools of resin can be found, the woodpecker keeps the sap flowing, and it is thought that this discourages predators such as snakes. They live in small groups of family members though there will be only one breeding pair among the group. Each group territory of around 80–120 ha will have cavities in a number of trees called a 'cluster'. Despite the name the red 'cockade' found just below the distinctive black cap is only visible in breeding males. Breeding takes place in late spring, the female moves into the male's cavity and lays three to four eggs. All members of the group help to incubate the eggs which will hatch after 10–12 days, then all the group members help to forage for the hatchlings until they fledge after about 26 days. The newly fledged juveniles swell the group size, though the young females often leave the group the following spring to find mates.

Photo: Red Cockaded Woodpecker © Jim Hanula of the US Forest Service

Conservation Status:

- IUCN Red List – Vulnerable Status
- Audubon Society Watchlist 2007 Red Status
- USFWS – Endangered

Current Status:

Indigenous to the US the once common woodpecker was found across the eastern seaboard and west to Tennessee and Eastern Texas. The Red Cockaded Woodpecker is now currently found only in 11 states. The USFWS is undertaking a major recovery plan associated with the long leaf pine. This includes protection status, excavation of cavities to be used by the woodpeckers and in some states private landowners are given small financial incentives to implement conservation measures. It should be noted that the conservation status of this bird is linked not just to the long leaf pine, but also to many other forest species of bird, small mammals, insects and reptiles that are dependent on residing in abandoned tree cavities.

References

Sibley, D.A. (2003) *The Sibley Field Guide to Birds of Eastern North America*, Alfred A
    Knopf Inc., © Chanticleer Press, Inc., New York.
www.fws.gov
www.audubon.org

Numerous initiatives have developed over the last decade and federal government has
initiated a series of programmes to aid both LLP restorations in public and private lands.
The US Army Corps of Engineers has a new ecosystem management strategy and is involved
in several restoration projects across the southeast including a 300-acre (121 ha) site
at Allatoona lake in Georgia. The list of agency involvement is impressive, ranging from
The Nature Conservancy, USDA Forest Service, US Department of Interior Fish and Wildlife
Service, US Department of Defence, Cooperative Extension Service, Tall Timbers Research
Station, Joseph W. Jones Ecological Research Centre and State agencies and numerous
private landowners. In 1995 the Longleaf Alliance was formed to serve as a forum for
dissemination of information.

The federal government offers financial aid to these agencies through a number of programmes including the CRP, Forestry Incentive Programme, Forest Stewardship Programme,
Partners with Wildlife Program and the Safe Harbour Program. Recently under the CRP a
modest expansion of LLP forest has been realised when for two years LLP was the most
frequently planted tree, indeed from 1998 to 2000 68 240 ha of LLP were planted across the
south eastern region (Brockway *et al.*, 2005).

LLP forests also occur across significant tracts of private land and during May 2008
we were treated to a guided tour by one private landowner who is making a significant
contribution to the conservation of LLP. Our guide was Kevin Williams of the USDA Forest
Service and we were shown around 600 acres (249.9 ha) of LLP forest at Kalawi Farm, Eagle
Springs, Moore County, North Carolina (NC). The farm is owned by Art Williams and family
and extends to 3000 acres (1214.6 ha). The farm has integrated LLP conservation into the
business structure of the farm and as such manages the LLP forest to produce bales of pine
needle mulch for the ornamental nursery trade and game cover; on our visit we saw several
wild turkeys.

Kalawi Farm is situated in the sandhills of NC and therefore the LLP forests in this region
are typically xeric forest and have a diverse range of plant and animal species (Table 4.10).
Managing these forests for conservation is challenging and, like many other LLP forest
types, there is significant encroachment of understorey and midstorey hardwoods.

There are three categories of degradation that can be observed in xeric LLP forests
(Table 4.11), which depend on the degree of hardwood in-fill, so each category has a range
of restoration principles (Table 4.11). The basic principle is to reintroduce prescribed burns
to remove or control the hardwood in-fill and other invasive canopy pines, in this case
sand pines. Where burning is not possible due to restrictions imposed by close proximity
to urban areas herbicide treatments (Hexazinone and Glyphosate based products such as
Roundup) can be used to eradicate hardwoods. Indeed herbicide applications in conjunction
with prescribe burning can speed up restoration work, especially where oak is a problem.
Application rates of 1–2 kg of active ingredient per hectare will produce 80–90% mortality

Table 4.11 Categories of degradation for sandhills longleaf pine and associated management strategy

	Moderately degraded	Very degraded	Highly degraded
Canopy trees	Longleaf pine	Other trees	Other trees
Understorey	Native plants	Native plants	Non-native plants
Xeric and subxeric sandhills	Growing season fire, dormant season fire, mechanical removal and herbicide treatment of hardwoods	Mechanical harvesting, growing season fire, herbicide sprouting hardwoods, plant LLP seedlings	Roller chop twice and burn, herbicide where necessary, plant LLP seedlings and sow native understorey seed

in oak with minimal long-term damage to the herbaceous understorey species (Brockway *et al.*, 2005).

When burning is being carried out, care must be taken to assess the depth and moisture of the fuel load. This is particularly pertinent where fire has been suppressed for many years leading to a deep litter load (Brockway *et al.*, 2005). The issue with deep litter in xeric stands, during the growing season, the litter layer is dry to the base of the tree, any subsequent fire will potentially damage feeder roots and the cambium at the root collar (Brockway *et al.*, 2005). Therefore if LLP stands have a deep litter layer, then dormant season burns would be a more suitable approach, as the litter layer will have retained some moisture, which will produce a cooler burn.

The action of burning opens up the understorey affording opportunities for re-colonisation of the bunchgrass and herbaceous plants as fire stimulates new grass growth from surviving basal meristems and flowering and seed set during the following season. Fire also stimulates dormant seed banks and consequently a flush of herbaceous plants. Another advantage of introducing prescribed fires is the control of fungal pathogens and suppression of insect pests.

Where LLP stands are seriously depleted of bunchgrass it is important to restock, and plugs of appropriate species can be planted out during spring at 1 m by 1 m centres and an application of fertiliser during the second or third growing season will improve grass growth. Seed sowing is possible between the rows of trees (Brockway *et al.*, 2005).

At Kalawi Farm one of the objectives is the production of needle mulch and raking the forest floor, collecting and bagging the litter, which produces 40 000 bales annually. This action helps to reduce litter build up and consequently maintaining healthy LLP stands via burning is safer due to reduced fuel loads. There is a trade-off. If the litter harvest is excessive and overdone then nutrient recycling is reduced and tree growth will be curtailed. Reductions in natural nutrient recycling can be offset by applications of super-phosphate at 224 kg per ha but this is expensive and the favoured option is to monitor the needle harvest.

Another objective of the LLP management at Kalawi Farm is to provide natural cover for game shooting, this has been combined with the field margins (Chapter 2) and consequently there is a healthy population of game such as quail, wild turkeys and deer, but also rare species such as the red-cockaded woodpecker and the fox squirrel (Species Box 4.4) which we observed on our visit.

Species Box 4.4: Fox Squirrel (*Sciurus niger*)

Profile:

The Fox Squirrel is named from its fox-like tail and can be found across the eastern and central US, southern Canada and northern Mexico. Although generally relatively common, squirrels with a very dark coloured variation are fairly rare and found only in the southeastern US. It is the largest of the tree squirrels, 45–70 cm in length with a 20–33 cm tail, and weighing in at between 500 and 1000 g. They are strictly diurnal and are opportunistic in their eating habits though their preferred foods are acorns, seeds, nuts and pine kernels. They, in turn, are preyed upon by birds of prey, snakes and bobcats. Males and females are alike (no sexual dimorphism) and they are promiscuous.

Photo: Fox Squirrel © Stephen Burchett 2008

Conservation Status:

- IUCN Red List – Least Concern Status
- US Federal List – Endangered
- Smithsonian Status – Near threatened

Current Status:

These southeastern fox squirrels live mainly in mature LLP forests. Their numbers have been reduced through loss of habitat, and their recovery is inexorably linked to the recovery and restoration of these forests.

Reference

Fahey, B. (2001) Sciurus niger online, animal diversity web. At http://animaldiversity. ummz.umich.edu (accessed 20 September 2009).

In summary, the management of LLP at Kalawi Farm has a firm financial underpinning. However, as noted by other authors (Brockway *et al.*, 2005; Alavalapati, Stainback and Carter, 2002) LLP forest on private lands, has in the past, been replaced by commercial pines to improve farm income, at Kalawi Farm this is not the case and an extensive system of healthy LLP forest form part of the landscape matrix.

Case Study 4.3: New Hampshire Woodlands

Urbanisation of America's Forests

A major threat to America's forest is caused by one underlying problem, population growth and urbanisation. Decline in the extent of and the degradation of LLP is one that has been driven by commercial exploitation from the direct use of the timber to complete conversion of the forest to agriculture and, in modern times, fragmentation due to urban sprawl. During the 1990s 7.8 ha (19 acres) per hour of forest, wetlands and agricultural land was urbanised (Vince *et al.*, 2005). The southern states are not the only states subject to urban sprawl; in the northeastern state of NH rapid urbanisation in the southern half of the state is threatening the ecological integrity of extensive areas of mixed northern forest. The key issue here is the decline in working woodlands and the subsequent impacts on the natural biology of the wildlife associated with these woodlands.

The forests of NH cover about 78% of the state, but in many parts of the state urban development has led to increased fragmentation and consequently reduced the extent of forest patches and ultimately their commercial value. This leads to a loss in regional investment in the infrastructure, which supports the commercial loggers. On the face of it this may appear beneficial to wildlife but this is not the case. Land use will change as people begin to realise that profit can be made by another type of land use, such as housing and urban developments. When the potential income from urbanisation is compared to the elusive non market values provided by forest cover (water catchments, wildlife refuge and carbon sequestration) then more often than not urbanisation will occur. The pressure to convert forest and indeed wetlands and agricultural land will increase as economic growth in cities brings new development, new residents and new demands for land use and ultimately urbanisation and hence forest fragmentation.

In NH, as the extent of forest cover shrinks, the commercial loggers lose interest in the smaller patches and withdraw financial investment in the region and hence the land becomes available to urban development as lots are sold off for private homes, often second homes. The critical threshold of forest size for commercial loggers in NH is around 200 ha (500 acres) and only 26% of remaining forests in NH exceed this size.

Increasingly forest conservation in NH is now taking place on the land occupied by smaller private NIFOs and, for many of these forest owners, logging and commercial sale of lumber is not a high priority. Nonetheless for the continued conservation of these northern mixed forests maintaining working woodlands is essential (Malin Clyde, Personal Communication 2008) and the University of New Hampshire Extension Service is working with forest owners to provide advice on forestry management. One such example is the Ames Road Forest Tree Farm, NH, owned by Ned Therrien. Ned's tree farm extends to 48.6 ha (120 acres) with 40.5 ha (100 acres) of forest and the remainder being occupied by wetlands and open areas. The composition of trees in the forest is typical of this area such as white pine (*Pinus strobes*), white birch (*Betula papyrifera*), sweet birch (*B lenta*), sugar maple (*Acer saccharum*), red maple (*A rubrum*), ash (*Fraxinus spp.*), cherry (*Prunus spp.*), red oak (*Quercus rubra*) and hickory (*Carya spp.*).

The main objective is the production of quality saw logs (timber) by encouraging the regeneration of white pine and red oak. Other activity at the site includes landscape and wildlife habitat conservation. The management of the woodland is divided into 58 permanent plots, where stock inventory is taken every 10 years. Ned purchased the land in 1980 and has been making considerable progress in conservation since. To help fund the conservation works there have been four commercial timber sales (1985, 2001, 2005 and 2006) since 1980 where the total volume of timber logged is in the region of 5419 m^3. This lumber has been divided into two classes; saw logs where a total of 505 400 board feet (1193 m^3) has been sold and firewood and pulp, where a total of 1166 cord (4226 m^3) has been produced. The sale of this resource helps to fund on site conservation work.

In terms of conservation effort the theme has been to reintroduce open areas, ponds and wetlands following the return of beavers. This work can actually dovetail into woodland management for commercial timber production. Open areas are an important conservation aim in NH (Chapter 3) as they provide a number of habitats for birds and mammals, but in the case of Ames Road Forest removing over-storey trees creates these important open gaps but also helps to promote the regeneration of white pine. Ned has removed 10 acres of over-storey trees with the aid of federal grants and has created a patchwork of open areas and simultaneously increased the number of white pine seedlings. Another management strategy for these clearings is to actually increase the size of the clearings and here Ned's aim is to increase the soil temperature to improve establishment of hardwood saplings. This is an important management practice because the understorey has a low density of seedlings due to the dense crowns of the canopy. The spin-off for wildlife in this long-term strategy is an increase in oaks, which will eventually provide more winter-feed for wildlife. The debris created from all this logging is left in piles providing important refuges for numerous animals.

Naturally, forest and woodlands are not just areas covered in trees but in reality are landscapes composed of many habitat niches and certainly in many forests throughout the world, standing water, vernal pools, marshes, rivers and many other small watercourses abound and in NH all these aquatic habitats were once common in woodlands across the state. Now these habitats, like the woodlands themselves are threatened by urbanisation.

Watercourses, open bodies of water and vernal pools are important conservation areas and Ned is keen to encourage and protect these habitats. On our visit we saw three large standing bodies of water and one major beaver dam (Figure 4.13) which has returned a low lying area of woodland back into a flooded plain. Ned commented that when he purchased the woodland in 1980 there were no flooded areas and these have only returned as the beavers started to dam up watercourses. These dammed areas are important to the woodland ecology and Ned has actively encouraged wood duck (*Aix sponsa*) and hooded mergansers (*Lophodytes cucullatus*) by placing nest boxes in the pools (Figure 4.14) and reports 214 wood ducks and mergansers have fledged from these pools since 1980.

Within Ned's tree farm there were many small pools that clearly constitute vernal pools. Vernal pools are small hollows, large scrapes and other depressions that hold water from autumn to early summer, typically for about 11 months in these NH woods. This cycle of wetting and drying is important and characterises the animal life associated with these features. Generally the amphibians such as salamanders, like the spotted salamander (*Ambystoma maculatum*) and the blue spotted salamander (*Ambystoma laterale*) and wood frogs (*Rana sylvatica*) are important elements of the fauna observed in these pools but other faunal groups also use the vernal pools as a habitat, examples include the fairy shrimp (*Anostraca*, numerous families), Blandings turtle (*Emydoidea blandingii*), great blue heron

Figure 4.13   A beaver dam impedes the natural flow of woodland stream leading to the development of a flooded plain. (Photo: Sarah Burchett)

Figure 4.14   Wood duck nest box. (Photo: Sarah Burchett)

(*Ardea herodias*) and numerous larvae of predatory invertebrates. Generally these pools are used as breeding grounds and once the aquatic side of the life cycle is complete these animals leave the pools and use the surrounding woodland for the rest of their life cycle, moving out from the pool to a distance of some 181 m (600 yards). Staff from UNH are monitoring these pools for aquatic life and were very pleased to report a healthy population of life including marbled salamanders (*Ambystoma opacum*). Ned's tree farm is also a patchwork of small plots with stonewall boundaries, a relic of past land use, agriculture that was abandoned during the era of the great gold rush. Adjacent to many of these boundaries is a network of ditches and while walking through the woodland we spotted several wood frogs (*Rana sylvatica*) basking in the warmth of the spring sun.

To ensure minimal damage to these important aquatic habitats Ned has developed a series of permanent extraction routes and staging yards, which reduce the environmental risk to these pools from logging skids, and soil erosion as a result of associated soil compaction caused by unplanned timber extraction routes. Many of these woodland tracks have been planted with a conservation mix of grasses and understorey herbaceous plants, providing yet another habitat, which threads throughout the woodland connecting one area to another and consequently acting as wildlife corridors. There is clearly a philosophy of sustainable forest management at Ned's tree farm and this is supported by extension officers from UNH and consequently the commercial sale of lumber from Ned's tree farm maintains these woods as working woodlands and hopefully staving off future urban encroachment. Indeed as we left Ned his last comment was that he is considering placing the whole woodland in a conservation easement, an action that would safeguard the site forever.

Case Study 4.4: Malaysian Tropical Forests, Forestry Industry and Enrichment Planting

The Malaysian forestry industry is primarily based on tropical hardwood from the dipterocarpaceae family of trees (Species Box 4.5) and a selection of other tropical hardwood trees, including the Ebenaceae and plantation teak. Also, once exhausted of latex old rubber trees are harvested. The extent of logging is huge, for example on the island of Borneo forest cover in the mid-1980s was still at 75% but by the turn of the twenty-first century forest cover has declined to 50% and continues to decline (Garbutt and Prrudente, 2006), with the average annual loss of forest cover estimated at 237 000 ha, which is largely attributed to land use change (ITTO, 2005). However, Garbutt and Prrudente (2006) give a figure of 850 000 ha per annum for Borneo.

Exploitation of these forests was initially under the Malaysian uniform system (MUS), which was established following 50 years of research (Contreras-Hermosilla, 1999). The MUS is a silviculture system that involves felling commercial trees of a diameter at breast height (dbh) of more than 45 cm on a 50–60 year cutting cycle. Following cutting forest workers would re-enter a coupe and cut out any non-commercial trees of 15 cm dbh, this was to reduce competition. In the early years of logging in Malaysia, the MUS was very successful

Species Box 4.5: Dipterocarpaceae

Profile:

A large pan tropical family divided into three distinct subfamilies; the Dipterocarpoideae, Monotoideae and the Pakaraimoideae. The dipterocarpoideae is the largest subfamily containing 13 genera and 475 species distributed from India to South East Asia with a large distribution in Malaysia, in particular Sabah and Sarawak. The name dipterocarpaceae is based on the first described genus the *Dipterocarpus*. This genus has a fruit that consists of two wings (dip = two, pteros = wing and carpos = fruit). However, not all the genera and thus species have two wings, for example *Shorea foxworthyi* has five wings.

Photo: Dipterocarp © Stephen Burchett 2008

Dipterocarps are tropical hard wood trees and consequently form the basis of the trade in tropical hardwoods. Many genera within the dipterocarpoideae have numerous species that form large emergent trees, for example S. *foxworthyi* extends to 60 m in height and has a 1.2 m diameter with a dense hemispherical crown and a very straight bole. Commonly found in lowland dipterocarp forest to about 700 m altitude.

Conservation Status:

- Currently endangered due to habitat loss and land conversion to oil palm

- Listed on IUCN Red List as critically endangered

Photo: Winged Dipterocarp seeds © Stephen Burchett 2008

- A favourite tree for loggers due to its height (60 m) and thus a major source of timber. Its reproductive cycle exceeds the present cutting cycle (IUCN, 2009).

in regenerating lowland dipterocarp forest (Contreras-Hermosilla, 1999) but, during the expansion of the country's agricultural sector and in particular oil palm, large tracts of logged-over forest were converted to agricultural use and the MUS system fell into disuse. Coupled with the expansion of logging into the hill forest, up to the 1000 m contour line, where the MUS was found to be less successful (Chan, Shamsudin and Ismail, 2008) the MUS method has largely been replaced by the Selective Management System (SMS). This method differs from the MUS in that the cutting cycle is reduced to 25–30 years, the minimum dbh is set at 30–50 cm and there is no follow up management of the non-commercial pioneer species. Another major drawback of the SMS is the use of tractor and high lead logging which causes significant damage to saplings and soil structure. When this is coupled with the extensive size of the logging coupes, leading to fragmentation and isolation of reproductive mother trees, and the rate of logging, which often exceeds the reproductive cycle of the dipterocarps, there is a significant decline in recruitment and regeneration of dipterocarps. Consequently there is a genuine threat to the biological integrity and long-term security of the dipterocarps. The level of endemism observed in the Malesian region further exacerbates this threat.

Status of Malaysian Forests

Malaysia has a population of 25 million people and a land area of 32.9 million ha, comprising 13 states. Within this landscape, estimates of total natural forest cover range from 19.3 million ha to 19.5 million ha (ITTO, 2005). The forest sector is divided into permanent forest estate (PFE) which is all natural forest and in 2003 covered 14.39 million ha (ITTO, 2005) and planted forest (PF) which is plantation forestry covering 263 000 ha in 2003 (ITTO, 2005). Within the PFE there is a total of 3.21 million ha of protection forest leaving 11.18 million ha of production forest, which is fully gazetted in accordance with the National Forest Act 1984 (ITTO, 2005). The rationale for gazetting is to secure the timber trade and allow surveillance and law enforcement, in particular to protect it from illegal logging.

There has been movement towards sustainable forest management and certification and within the PFE 6 790 000 ha of forest have been allocated to concessions under license, of which 4 620 000 ha are certified (ITTO, 2005).

Lowland Mixed Rainforest

The term lowland mixed rainforest refers to both deciduous and evergreen mixed rainforest that occur across the tropical regions from sea level up to 1200 m. These forests are unquestionably some of the most diverse ecosystems in the world and islands such as Borneo are recognised by the international scientific community as biodiversity hotspots. In context with the Malaysian enrichment planting projects the focus here will be on lowland forest dominated by one family of trees the dipterocarpaceae (Species Box 4.6). The presence of the dipterocarp trees is very important as it is this one family of trees that underpins the world's trade in tropical hardwood and therefore results in significant large scale commercial logging. In keeping with the biological and commercial importance of the dipterocarp trees from this point onwards, lowland mixed rainforest will be referred to as lowland dipterocarp forest.

The Malaysian lowland dipterocarp forests are dispersed across peninsular Malaysia and the states of Sabah and Sarawak in east Malaysia on the island of Borneo. The diversity of life in these forests is exceptionally high, for example on Borneo there are 2500 tree species with the number of dipterocarp species recorded at 267 species in 9 genera (Soepadmo, Saw and Chung, 2004). This staggering diversity extends to many other floristic families and the Orchidaceae is well represented in both Sabah and Sarawak. It is estimated that there are between 2500 and 3000 species of orchid, many of which occur as canopy epiphytes (Chan et al., 1994). In total, across the Malaysian region it is estimated that there are 25 000

Species Box 4.6: Bornean Clouded Leopard (*Neofelis diardi ssp. borneensis*)

Profile:

A secretive and rare cat which is dependent on the large tropical forested areas of Borneo. They are found mainly in virgin forest and sometimes extend their territories into nearby logged-over forest. Very little is known about them. They are thought to be solitary except for breeding, with an estimated density of between six and nine individuals per 100 m². Males are approximately 50% larger than females and their overall size ranges between 60 and 75 cm at the shoulder, 15–25 kg in weight and 140–180 cm in length,

Photo: Clouded Leopard in Borneo © Stephen Burchett 2006

of which half is the tail. Expert climbers, they are arboreal and terrestrial hunters, feeding on monkeys, deer, bearded pigs, birds and rats. They have very sharp claws for gripping and rotating rear ankles for dexterity in the trees and allowing them to descend a tree head-first.

Conservation Status:

1. IUCN Red List Status – Endangered 2008.
2. Endangered under the US Endangered Species Act.

Current Status:

Once thought to be part of a larger group of clouded leopards, the Borneo Clouded Leopard was re-categorised in 2007 as belonging to a separate sub-species. As such, these sightings were an extremely rare privilege and

Photo: Clouded Leopard Cub in Borneo © Stephen Burchett 2006

numbers are currently unknown though estimates range from a little over 1000 to in excess of 3000 individuals. The endangered category is in place as the Borneo Rainforest – the leopard's territory – is seriously under threat. As a consequence of the re-classification, a number of research projects are being initiated in Borneo to study the cats.

References

Garbutt, N. and Prrudente, C. (2006) *Wild Borneo. The wildlife and scenery of Sabah, Sarawak, Brunei and Kalimantan*, New Holland, London.
www.nationalgeographic.com

species of flowering plants. This complexity of life is related to the structural diversity of these forests, and within the lowland dipterocarp forests there are four recognised canopy layers:

1. The emergent layer from 35 m+ with sympodial crowns
2. The main canopy between 20 and 30 m
3. The understorey up to 20 m with monopodial crowns
4. The shrub layer.

Figure 4.15 A *Koompassia excelsa*, the local name for this species is menggaris, and many individuals were left standing after the first logging operation, as the wood from these trees is exceptionally dense and shatters on hitting the ground thus negating any commercial value. These trees can grow to 80+ m in height and form a distinct emergent canopy. (Photo: Stephen Burchett)

The emergent layer is a discontinuous layer that extends from about 35 to 95 m. Trees occur as isolated individuals or in small groups and are predominantly composed of dipterocarps; indeed up to 70% of the emergent species in this layer are dipterocarps (Soepadmo, Saw and Chung, 2004). However there is another important emergent, commonly known as the mengaris tree (*Koompassia excelsa*), that can reach up to 88+ m (Figure 4.15). Unlike the dipterocarps Mengaris trees do not have the same commercial value and consequently in logged-over coupes in Sabah many Mengaris trees stand as isolated remnants of the original emergent canopy. Emergent trees are characterised by large straight boles and have enormous buttresses for support (Figure 4.16). There are a limited number of canopy epiphytes found in this layer, due to extreme abiotic environment, and climbers are limited to the bole with the exception of strangler figs (*Ficus spp.*). When one of these canopy giants is felled through natural process (Figure 4.17) a gap is generated and a new cycle of forest regeneration begins.

Figure 4.16 Giant buttress of a *Koompassia excelsa*, research assistant from Danum valley conservation area, preparing to climb to the canopy, illustrates the scale of these structures. (Photo: Stephen Burchett)

The next layer is the main canopy and this is a continuous layer that can range from 20 m to about 30 m, depending on the location. There are a number of tree families represented in this layer, with important groups being the Dipterocarpaceae, Dilleniaceae, Ebenaceae, Euphorbiaceae, Fagaceae, Lauraceae and Leguminosae (Lee, 2003). Trees from the *Macaranga* genus are members of the Euphorbiaceae family and form the major pioneer species in logged-over forest. This layer has a significant epiphyte community which includes ferns

Figure 4.17 Falling, giant, emergent trees create gaps and opportunities for forest regeneration. (Photo: Stephen Burchett)

such as the birds nest fern (*Asplenium nidus*), orchids with examples from the following genera; *Bulbophyllum, Dendrobium, Dendrochilum, Laelle, Luisia, Porrohachis* and *Vanilla*. Connectivity between this layer and the other three layers occurs via an intricate network of climbers. Many fruiting trees can be found in this layer and as a result this is the home of the orang utan (*Pongo pygmaeus*) and other frugiverous mammals and primates. In turn, this layer attracts carnivores, including snakes but also mammals such as the civet cat (*Hemigalus spp.*) and the enigmatic clouded leopard (*Neofelis diardi subsp borneensis*), which are arboreal hunters (Species Box 4.6). The clouded leopard is listed as vulnerable on the IUCN red data list.

Below the main canopy is the understorey, which is composed of trees below 20 m and contains many young trees that will eventually become the emergent canopy. Here young emerging canopy epiphytes such as the birds nest fern (Figure 4.18) can be observed, these young ferns are the origins of the large specimens observed in the main and emergent canopy. It is this canopy that responds to gaps generated by falling emergent trees. Young trees will rapidly respond to the new light levels and extend to the main canopy and can eventually become emergent trees. The saplings in this layer have very slow growth rates and often reside in this layer between 10 and 15 years before their next phase of growth. These saplings also have exceptional resilience to invertebrate herbivory due to the high levels of toxic secondary metabolites in their leaves. An inspection of the foliage of saplings in this layer will quickly illustrate this point as very few present leaves showing symptoms of invertebrate herbivory (Figure 4.19).

Figure 4.18 Epiphytes such as these two large birds nest ferns (*Asplenium nidus*) in the main canopy layer provide opportunities for invertebrate colonisation, particularly within the decaying mass below the live fronds. (Photo: Stephen Burchett)

Figure 4.19 Many tropical trees produce an array of secondary metabolites that have pesticide like activity, preventing invertebrate herbivory and increasing probability of long-term survival in the understorey. (Photo: Stephen Burchett)

The final layer is the shrub layer and technically this layer can be viewed as understorey vegetation. However, in the context of lowland dipterocarp forest, the shrub layer is an important distinction. In this layer important groups of plants can be found that form extensive stands of pioneer plants in logged-over forest. Examples include species from the Palmaceae, Euphorbiaceae, Rubiaceae and the Zingiberaceae (ginger lilies). In disturbed forest the shrub layer can be dominated by species from the ginger (zingberaceae) and palm families, Palmaceae. Indeed there are 288 species in the zingberaceae family.

It is this structural complexity that underpins the huge diversity in life. The matrix of habitats within one hectare of lowland dipterocarp forest is substantial, ranging from exposed outer limbs of emergent trees, to moss infested branches in the main canopy, fruiting vines and rotting boles, standing and fallen deadwood all contribute to niche differentiation. Consequently life naturally exploits these opportunities. A study of invertebrate life will illustrate this point.

Invertebrate diversity in these forests is overwhelming and most likely will never be fully documented. In five commonly recognised groups estimates on species diversity are as follows:

- Ephemeroptera (Mayflies) = 50–100
- Odonata (Dragonflies) = 180
- Phasmida (Stick and Leaf insects) = 300
- Coleoptera (Beetles) = 30 000+
- Lepidoptera
  - Butterflies = 1000
  - Moths = 20 000+ (Hill and Abang, 2005).

The above are estimates derived from known species described across the South East Asian region (Hill and Abang, 2005) but are considered as robust indicators of the diversity in invertebrate life. However these are common groups and although form important ecosystem functions within the lowland dipterocarp forest they are by no means exclusive. Important groups, such as the termites (Isoptera), number 2300 species and their contribution to nutrient recycling is probably immeasurable but without this group natural forest regeneration would be impossible. Another important invertebrate group is the hymenoptera, which is composed of the ants, wasps and bees, here species diversity is estimated in the region of 100 000+ (Hill and Abang, 2005). Again, their contribution to ecosystem function is immeasurable. Ants, for example, species from the *Camponotus* genus, can be observed on the forest floor and in the canopy of many emergent trees, including personal observation in *Parashorea malaanonan* a giant emergent capable of heights of 60 m and having a girth of 2 m (Soepadmo, Saw and Chung, 2004).

The above discussions on biodiversity and plant-animal interactions represent a very limited introduction but illustrate one very salient point. Logging and forest conversion lead to changes in forest structure and assemblages of the vegetation which will have a profound impact on global biodiversity and consequently global ecosystem function. Continued unrelenting exploitation of these forests is unacceptable and pessimistically likely to continue due to humanity's insatiable reproductive appetite. In an attempt to readdress the continued loss of lowland dipterocarp forest the Malaysian government, in partnership with their research institutions and foreign sponsors has initiated several forest enrichment programmes.

Malaysian Enrichment Planting Projects

The original concept of enrichment planting was aimed at increasing the stock of commercial tree species following logging operations. This was essential as the logging rotation is often shorter than the reproductive cycle of the numerous dipterocarp species. Enrichment planting is targeted at forests that have been degraded by logging and agricultural conversion and there are three types of degraded forests recognised by the ITTO (2002):

- **Degraded forests:** A degraded primary forest is a primary forest in which the initial cover has been adversely affected by the unsustainable harvesting of wood and/or non-wood forest products so that its structure, processes, functions and dynamics are altered beyond the short-term resilience of the ecosystem; that is, the capacity of the forest to fully recover from exploitation in the near to medium term has been compromised.

- **Secondary forests:** A secondary forest comprises woody vegetation regrowing on land that was largely cleared of its original forest cover (i.e. carried less than 10% of the original forest cover). Secondary forests commonly develop naturally on land abandoned after shifting cultivation, settled agriculture, pasture or failed tree plantations.

- **Degraded forest land:** Degraded forest land is former forest land severely damaged by the excessive harvesting of wood and/or non-wood forest products, poor management, repeated fire, grazing or other disturbances or land-uses that damage soil and vegetation to a degree that inhibits or severely delays the re-establishment of forest after abandonment.

The selection of appropriate species for enrichment planting has been steered by ITTO guidelines on restoration, management and rehabilitation of degraded secondary forest (ITTO, 2002). The criteria for selecting species include:

- Species must be easy to handle in the nursery
- Species must have a good germination rate
- There needs to be good biological knowledge on the phenology of flowering and fruiting, with priority on regular flowering and fruiting
- High initial growth rate
- Shade tolerance
- Good competitive ability over other vegetation
- Good self pruning
- Low susceptibility to disease, insect and fungal attacks.

Enrichment Planting in Borneo

In the state of Sabah on the large tropical island of Borneo there are several substantial forest enrichment planting projects taking place. The aim of these projects is to improve biological diversity, contribute to the continuation and diversity of commercial tropical timber trees and to enhance the structure of severely degraded logged-over, permanent forest. Funding for these enrichment planting projects comes from a number of sources including commercial European companies, and partnerships are developed between the funding companies and the forest management company, Yayasan Sabah, who manage a 1 million ha forestry concession.

The IKEA Furniture Company and Innoprise Corporation Sdn. Bhd., a subsidiary of the Yayasan Sabah Foundation, funds one project, the INIKEA project. A memorandum of Agreement was signed between the Swedish sow-a-seed foundation and the Innoprise foundation of Sabah in 1998. The Swedish University of Agricultural Sciences supervises the project. Another very important enrichment planting project is the Innoprise- forest absorbing carbon dioxide emissions (FACEs) Foundation of the Netherlands carbon sequestration project in the Ulu Segama Forest Reserve near Lahad Datu, Sabah, Borneo.

INIKEA Project: Biodiversity Improvement of Degraded Forest in Sabah

The primary aim of the INIKEA project is to improve the biodiversity of seriously degraded tropical rainforest enabling the migration of natural flora and fauna back into the rehabilitated area. The project area is a contiguous block of heavily degraded forest in the Kalabakan Forest Reserve, west of Tawau. Degradation of the native forest occurred through wildfires in 1983 and shortly afterwards logging further compounded the problem. The resulting secondary forest contains compacted and degraded soils, and is dominated by climbers and pioneer species, typically from the Euphorbiaceae family.

Owing to the low number of mature mother trees remaining in a logged-over forest, the thick dense growth of climbers and the heavily compacted soils, recruitment of a new population of dipterocarp species is very poor. The dense canopy of the pioneer species further compounds this. Predictably the canopy species are short-lived *Macarangas* such as *M. gigantifolia* and *M. triloba* but they cast enough shade and contribute significantly to interspecific competition, such that recruitment and establishment of dipterocarp seedlings is further impeded and natural seedling establishment is very poor or non-existent. Therefore severely-degraded forest is targeted for enrichment planting.

The project area extends to 14 300 ha and the initial phase aimed to rehabilitate 4000–5000 ha of degraded forest (Wan Mohd Shukri *et al.*, 2008). By February 2005 about 6100 ha of degraded forest had been enriched.

Nursery Establishment and Secure Supply of Indigenous Species

Two methods were applied to enrichment planting in the INIKEA project (i) line planting where new plants are planted at 3 m intervals along 2 m wide transects that are 10 m apart (Figure 4.20) and (ii) gap cluster planting where new trees are planted in small groups, usually groups of three, within a 10 m × 10 m subplot.

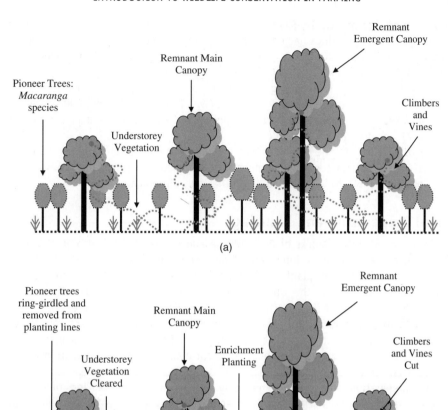

Figure 4.20 (a) Schematic representation of a logged-over lowland dipterocarp forest, illustrating the salient structural features prior to ground preparation for enrichment plantings. (b) Schematic representation of line planting, illustrating ground preparation, planting distances and impact of ring-girdling of pioneers trees and vine cutting

Before any of these techniques could be implemented it was essential to secure a source of seedlings. Consequently a nursery was established within the region to supply 500 000 seedlings per year. Nursery establishment required a supply of wild native seeds. To achieve this it was essential to survey and label wild mother trees, train and mobilise teams of seed collectors, with an emphasis on collection during mast events, which are notoriously unpredictable. The INIKEA project called upon the experiences of the Infrapro team (described below) to help them establish the seedling nursery (Falck *et al.*, 2007).

However, as dipterocarp masting is variable the project developed a process of collecting and revitalising wild seedlings named 'wildings' which provided an alternative to seed borne propagules (Falck *et al.*, 2007). All potted wildings were acclimatised in shaded humidity chambers (Wan Mohd Shukri *et al.*, 2008). Seedlings were then raised in either 7.5 × 20 cm or 15 × 22.5 cm black polythene bags, depending on fruit size, in amended forest topsoil, which had river sand added to improve drainage (Wan Mohd Shukri *et al.*, 2008). The INIKEA project imposed a compulsory no herbicide policy and thus in the nursery, a thick black plastic sheet was used as a weed mulch.

In the INIKEA project the emphasis was on habitat enhancement and thus only native seedlings were used and a species composition was made up from local trees, with 75% being dipterocarps, 5% fruit trees and 20% other non-dipterocarps.

Ground Preparation and Planting

The site was divided into blocks of 200–300 ha and further divided into 10–50 ha plots, which became the basic operational unit (Wan Mohd Shukri *et al.*, 2008). Vehicle access is essential for transporting plant stock and forest workers for routine maintenance and, following initial surveys of the operational units, new roads were constructed, where necessary, but a number of the planting sites just required an upgrade of ex-logging roads. In order to achieve the successful rehabilitation of the severely degraded forest, extensive groundwork was required to prepare the planting site. This was carried out manually using local labour and forest workers were required to cut planting lines and prepare grids for the new seedlings.

Ground preparations for the line planting started with the clearing of small non-commercial trees, shrubs and undergrowth across a 2 m wide strip for the length of the planting transect, which could be up to 500 m. Planting points were also marked out at this time. The next stage was a 100% cut of all climbers, followed by selectively ring-girdling any *Macaranga* trees along the planting lines. The overall approach to site preparation aims for a slow opening up of the canopy; this reduces soil evaporation, which subsequently helps to maintain soil health. The process begins with the clearing of the lower canopy, as described above, and the subsequent drying out of the cut climbers, this drying out process takes about two weeks (Wan Mohd Shukri *et al.*, 2008). The next stage of canopy opening lasts between 3 and 12 months depending on the time it takes for the canopy of the ring-griddled *Macaranga* trees to die back. Once all the ground preparations are complete and the canopy along the planting lines is open, planting of the enrichment species must take place immediately, otherwise the cleared lines will in-fill and spoil the site. As planting takes place each plant is given a 4 g per plant application of Agroblen fertiliser, a slow release fertiliser.

The other planting method used in the INIKEA project is gap-cluster planting, which was adopted in order to simulate natural gaps that occur in primary forest due to tree fall. In the context of enrichment planting the current philosophy is that gap-cluster planting allows greater flexibility in locating the planting points (Wan Mohd Shukri *et al.*, 2008). Establishing the quadrats within the forest necessitated establishing a 20 m × 20 m grid by cutting systematic lines 20 m apart and subsequently marking out each grid at 20 m intervals, which indicated the quadrat centre. These 20 m lines would then form the access and reference points. Once the 20 × 20 m quadrat was established 10 × 10 m subplots were laid out and within these 10 × 10 m subplots planting sites were selected based on site survey, where if less then five tree species were present planting would take place. Within the 10 × 10 m subplot a 3–4 m diameter planting gap was established and then the planting site is cleared of undesirable trees via ring girdling coupled with the clearing of understorey vegetation. Climbers were cut across the whole of the planting area. Once the site is finally prepared a cluster of three enrichment seedlings were planted 1–1.5 m from the gap centre. About two thirds of the rehabilitated area has been planted using the gap-cluster method and one third using the line planting approach (Falck *et al.*, 2007) and a total of 42 dipterocarp species, 8 non-dipterocarps and 20 species of wild fruit trees have been planted (Falck *et al.*, 2007).

Maintenance and Survival

On going maintenance is essential, weeding and vine cutting must be carried out three times a year for the first two years and then it can be reduced in subsequent years as the saplings mature and grow away. After about two years post-planting shade manipulation is required which involves the removal of the upper canopy trees such as the *Macaranga* species. While this operation would open up the canopy it would simultaneously stimulate the growth of the understorey vegetation resulting in significant competition for resources. Thus in subsequent operations, about four years post-planting, the understorey vegetation (palms, bamboo, ginger lilies and small non commercial trees) are cleared to liberate the planted trees (Wan Mohd Shukri *et al.*, 2008). The ongoing maintenance programme is designed to last 10 years.

Reported survival rates are variable and are, in some cases, quite low. In a 100% field census on three-year-old blocks, mean survival rates for gap-cluster were recorded at 60 and 63% for the line planting (Falck *et al.*, 2007). Another census, which focused on four-year-old plots dominated by dipterocarp species, there was a reported 86% survival rate for line planting and 68% for gap-cluster and no significant differences were found for height and girth (Falck *et al.*, 2007). Chan, Shamsudin and Ismail (2008) reported survival being between 62 and 76% after six years post planting but where sites were very open and the ground flora dominated with *Imperata* grass, survival dropped to 62%. In all cases where survival dropped below 65% then restocking and or beating up operations were implemented (Falck *et al.*, 2007).

Sapling mortality was caused by a number of factors such as disease, invertebrate herbivory and drought. Drought is by far the most important variable and experience gained by the INIKEA team identified those planting operations which took place during the drought seasons of January to February and July to August resulted in a high incidence of mortality (Falck *et al.*, 2007).

The cost of rehabilitating 4700 ha of rain forest exceeded US$3 million equating to approximately US$687 per ha of planting. Road construction and maintenance was the highest financial burden at around US$668 000. During the wet season roads and bridges are often washed away (Figure 4.17). Administration and wages were the next largest costs totalling US$311 407. These outgoings were entirely based on local labour and profound socio-economic benefits were gained for the local population of forest workers and their families. A full account of these gains is described in Falck *et al.* (2007).

Implications and Global Benefits of Enrichment Planting

The INIKEA project is just one example of numerous enrichment programmes being imple-mented across logged over forest in both peninsular Malaysia and the states of Sabah and Sarawak in Borneo. The advances achieved by these programmes come from the close working relationships established between research institutes such as Forest Research Institute Malaysia (FRIM), the FACE of the Netherlands working with INFRAPRO and the Royal Society under the South East Asia Research Programme. One major issue with forest enrichment is a secure supply of dipterocarp seeds and seedlings, for example the INIKEA project used over a million seedlings during the planting phase of the project.

Flowering and fruiting in the dipterocarps is notoriously variable and in recent decades the well-known phenomenon of mast flowering (*en mass* flowering) has been signif-icantly perturbed. The mechanism of mast flowering and recent perturbations is not fully described but within the context of sustainable forest management, the Malaysian authorities, through FRIM and partners, have developed a database on dipterocarp and non-dipterocarp phenology. These data were derived from phenological studies conducted in seed production areas across peninsular Malaysia (Shukor, 2001). The phenological database now forms part of a wider database on forest genetic resources, the Forest Genetic Resources Information System (FOGRIS), which has been used across Malaysia to store information on seed production, seed storage (very difficult in dipterocarps) and location of mother trees. Predictions of seed production have been made using FOGRIS and the following general information helps enrichment project managers to secure seed supply;

1. Mast flowering season to fruiting lasts four to five months

2. Two distinct categories of flowering trees are identified

   (a) Species restricted to mast flowering

   (b) Species that flower regularly, either annually or biannually, but may skip one or two years or develop periods of irregularity in phenological behaviour

3. The majority of dipterocarps are confined to the mast flowering category where up to 70% of species will flower *en mass*

4. *Shorea* flower on an annual basis but with varying intensities (Shukor, 2001).

The complex reproductive biology of the dipterocarps coupled with their recalcitrant nature and with regards to long-term storage, can hinder enrichment projects and consequently FRIM has been developing protocols for clonal vegetative culture, tissue culture and cryo-preservation. To date success has been limited but the following species have been successfully cultured using vegetative cutting culture; *Shorea bracteolate, S ovalis, S parvifolia, S acuminata, S leprosula, S roxburghii* and *Hopea odorata* (Ahmad, 2006), and with tissue culture methods; *S leprosula*, and *S roxburgii* (Nakamura, 2006).

Cryo-preservation is a complex subject and a full account of the topic is beyond the remit of this book, however, the technique has considerable applications in conservation biology and thus merits some discussion here. Cryo-preservation is the long-term storage of germplasm at $-196\,°C$, or liquid nitrogen temperatures. The application of cryo-preservation is an iterative advancement of the general science of seed storage and three categories of seed tissue have been described as follows:

- Orthodox seeds are seeds that can be dried to about 5% moisture content without undue damage to their physiological and biochemical mechanisms. Seeds from this category can be stored for many decades or even longer providing no external fungal and bacterial pathogens compromise their biological integrity.

- Recalcitrant seeds are seeds from a range of plant species, typically from the tropics, that cannot withstand being dried below 26% moisture content and in many species this figure can exceed 50%. The following dipterocarps, *Shorea singkawang, S. pachyphylla* require a moisture content above 55% to maintain viability and *Vatica umbonata* will be irreversibly damaged if intercellular water content drops below 74% (Tompsett, 1998). Recalcitrant seeds will be damaged if they are stored at sub-zero temperatures and again in many cases will not tolerate temperatures below 16 °C (Krishnapillay and Tompsett, 1998).

- Intermediate seeds can be stored at much lower moisture contents than recalcitrant seeds, typically between 8 and 12% moisture content without significant loss of viability. This category of seeds can withstand some chilling and sub-zero temperatures, often by supercooling, but not to the same degree as orthodox seeds. When the physiology of these seeds were studied in detail it was found they had a high degree of similarity in their physiological and biochemical mechanisms as orthodox seeds but their desiccation tolerance had limits (Krishnapillay and Tompsett, 1998) and are often reported as orthodox with limited desiccation ability (OLDA).

There has been some limited success in the application of cryo-preservation in storing dipterocarp seeds such as *Dipterocarpus alatus* and *D. intricatus* (Krishnapillay and Tompsett, 1998), but these are orthodox seeds and consequently conventional seed storage technology is adequate. Cryo-preservation has been attempted with dissected embryos of *Hopea odorata* and *Dryobalanops aromatica*, but generally this has not resulted in any significant success (Krishnapillay and Tompsett, 1998). The main problem is the susceptibility of these seeds to desiccation damage, which can occur where intracellular freezing is initiated, such as during the freezing and thawing process used during a cryo-preservation protocol.

The above discussions illustrate some of the substantial difficulties facing enrichment projects and a full discussion on the ecology and biology of dipterocarp seeds and seedlings is given in Appanah and Turnbull (1998), with further examples in advancements in forestry technology outlined in Suzuki *et al.* (2006). Carbon trading will undoubtedly drive the emphasis on future enrichment programmes in South East Asia and CO_2 offset for Annex 1 countries such as the USA, Australia, New Zealand, Europe and the UK, has already been established in Sabah (Wan Mohd Shukri *et al.*, 2008).

Under the Kyoto protocol of the United Nations Framework Convention on Climate Change, industrialised or Annex 1 countries can meet part of their greenhouse gas emissions through carbon sequestration projects under the Clean Development Mechanism. This mechanism allows Annex 1 countries to invest in forest restoration projects such as enrichment planting. One such project is a partnership between Innoprise Corporation Sdn. Bhd and FACE (INFAPRO), which was established in 1992 in the Ulu Segama Forest Reserve near Lahad Datu (Map 4.4). The Ulu Segama reserve is adjacent to the Danum Valley Conservation Area

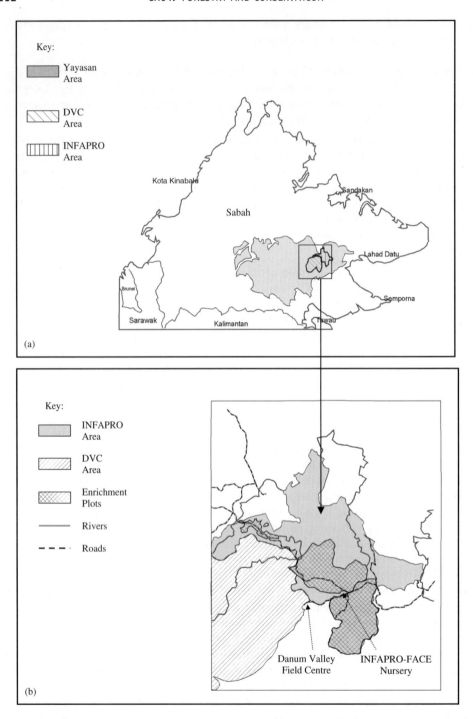

Map 4.4 (a) Location of Yayasan concession, Danum valley conservation area and INFAPRO project area in the state of Sabah. (b) Enrichment planting areas in relation to Danum valley conservation area

(DVCA) that covers 438 km^2 and is an excellent example of remaining lowland dipterocarp forest.

The objective of the INFAPRO project is to rehabilitate 25 000 ha of heavily logged-over rainforest, where logging occurred during the intense and damaging practices of the 1960–1980s. The forests in this region suffered significant detrimental impacts with respect to its structural complexity and as a result of high-lead or skidder extraction technology the soils were heavily compacted, thus ensuring significant impediments to natural regeneration (Reynolds, Personal Communication 2006). Up until 2009, 11 000 ha have been enriched using the line planting approach and 16 species of dipterocarps are commonly planted, including species from the following genera; *Dipterocarpus, Dryobalanops, Hopea Shorea and Parashorea spp.* (Anon, 2009). To date the estimated carbon held in the enrichment forest is 145.7 $\pm$ 9 tonnes/ha^{-1} and in the surrounding logged-over forest the carbon stock is estimated to be 92.5 $\pm$ 12.5 tonnes/ha^{-1} (Anon, 2009).

Although these projects have been running for the last 15 years or so there is currently little data on the benefits of enrichment planting for wildlife. Several research projects are ongoing and data should be published in the next few years. However, during several visits to the INFAPRO project a number of charismatic species were observed from the logging trails including the Asian elephant, orangutan, clouded leopard, several nests of Asian honeybees, civet cats, numerous bird species including the rhinoceros hornbill (*Buceros rhinoceros*) and the Asian pied hornbill (*Antbracocerus albirostris*). If these species are able to forage or reside within the project area it is a good sign that enrichment planting is improving habitat quality and thus resource availability for forest wildlife.

Planting dipterocarps also improves the diversity and population of this important family of tropical trees and when this is coupled with advances in the science of ex-situ and in-situ conservation biology the future of these trees, if not fully guaranteed, is clearly enhanced. One critical comment is that the diversity of dipterocarps being planted in these extensive enrichment areas is low. An example of dipterocarp diversity in an undisturbed primary forest is illustrated by one 50 m × 50 m quadrat. During one field visit in 2006 to a permanent quadrat on the west trail within the DVCA primary forest, we observed 16 mature species (d.b.h > 10 cm) of dipterocarps (Table 4.12) and in three years of field visits a total of 30 species have been recorded along the nature trail and the west trail. Clearly the total diversity of dipterocarps is important and therefore enrichment programmes should consider expanding their species list. This comment is quite pertinent when the total ecology of the forest habitat is considered; in particular, questions need to be addressed with respect to canopy ecology.

There is scant research being conducted on the epiphytic flora and subsequent colonisation of enriched forest and currently our knowledge of species diversity and structural complexity is limited. A small study looking at the epiphytic orchids in the canopy of five *Parashorea malaanonan*, within the DVCA during 2008 recoded 42 orchids from eight genera (Table 4.13) and one species which could not be identified from the current four volumes describing the Orchids of Borneo (O' Malley, 2009). Another common epiphyte in these forests is the birds nest fern (*Asplenium nidus*), which is commonly seen in the high canopy of primary forest and can reach an incredible size with mature specimens weighing 200 kg fresh weight. Canopy epiphytes of this size form excellent habitats for numerous species and there is a

Table 4.12 Dipterocarp species observed in a 50 m by 50 m quadrat in Danum valley conservation area

Scientific name	Local name	Height	IUCN red list conservation status
Dipterocarpus gracilis	Keruing kesat	45 m Emergent	CR
Dryobalanops lanceolata	Kapur paji	80 m Emergent	EN
Hopea nervosa	Selangan jangkang	30 m Canopy	CR
Hopea ferruginea	Selangan mata kucing	40 m Mid Canopy	CR
Parashorea malaanonan	Urat mata daun licin	60 m Emergent	CR
Parashorea tomentella	Urat mata beludu	65 m Emergent	Not listed
Shorea argentifolia	Seraya daun mas	45 m Emergent	EN
Shorea falciferoides	Selangan batu laut	50 m Emergent	CR
Shorea fallax	Seraya duan kasar	50 m Emergent	Not listed
Shorea johorensis	Seraya majau	50 m Emergent	CR
Shorea leprosula	Seraya tembaga	60 m Emergent	EN
Shorea parvifolia	Seraya punai	65 m Emergent	Not listed
Shorea pauciflora	Oba suluk	60 m Emergent	EN
Shorea superba	Selangan batu daun halus	75 m Emergent	CR
Shorea symingtonii	Melapi kuning	50 m Emergent	CR
Vatica dulitensis	Resak bukit	25 m Canopy	Not listed
Vatica sarawakensis	Resak Sarawak	30 m Canopy	CR

Note: CR = Critically Endangered, EN = Endangered.

Table 4.13 Orchid families associated with *Parashorea malaanonan*

Genera	Morpho species	Number of specimens
Bulbophyllum	6	24
Dendrobium	3	5
Dendrochilum	2	4
Laelle	1	1
Luisia	1	1
Porrohachis	1	3
Vanilla	1	3
Unknown	1	1
Total	16	42

clear exponential relationship between fern size and animal biomass ($R^2 = 0.96$) and on average these mature ferns can contain $88 \pm 14\,g$ of invertebrate biomass (Ellwood and Foster, 2004). It is essential for the long term durability and biological enhancement of these numerous enrichment projects that research is initiated to establish how canopies recruit their epiphytes and what impacts on total biodiversity planting a limited number of trees is going to have.

4.12   Summary

The literature is replete with discussions and examples of the increasing fragility and fragmentation of the world's forests and clearly human population growth is directly correlated with deforestation. When these discussions are reviewed in context with predictions in future population growth it becomes increasingly clear that the remaining natural tracts of forest will continue to be threatened, and this is not restricted to deforestation in the tropics. In the US, Europe and the UK urbanisation is another major threat, because as economic growth in developed countries continues to grow, cities and towns expand, road and rail networks bisect landscapes and fragment fragile woodlands leading to further ecological and rural degradation. The case studies presented above demonstrate that forest conservation by individual landowners; large-scale restoration initiatives and enrichment planting can be achieved leading to benefits for wildlife. The key issue in all these case studies though is that there must be a financial incentive. In the SOI the economic and social driver is aimed at improving the financial sustainability of local crofters and the wider community and if this is achieved then the Atlantic Oakwoods will continue to be managed by coppicing and consequently this will lead to diversity in habitats and landscape. In Malaysia the original objective of enrichment planting was to replace commercial timber trees but carbon offset projects are going to dominate the funding of future programmes. Again, the local community benefits through employment for forest workers. In the US, forest conservation and restoration has employed similar commercial overtones but generally it is controlled by the private sector and supported by government initiatives from Conservation Reserve Program (CRP) to advisory from USDA extension service. In all these case studies the balance between commercial exploitation and conservation is a fine one and continued adherence to SFM techniques is essential to maintain a viable forest for future generations.

References

Achnaha Community Woodland Management Plan (2007) http://morverncommunity woodlands.org.uk/publications/AchnahaCommunityWoodManagementPlan%2007-2012.pdf (accessed 24 June 2009).

Akeroyd, J. (2002) *The Encyclopaedia of Wild Flowers*, Paragon, Bath.

Ahmad, D.H. (2006) Vegetative propagation of dipterocarp species by stem cuttings using a very simple technique, in *Plantation Technology in Tropical Forest Science* (eds K. Suzuki, K. Ishii, S. Sakurai and S. Sasaki), Springer, Tokyo. Pp, 69–76.

Alavalapati, J.R.R., Stainback, G.A. and Carter, D.R. (2002) Restoration of longleaf pine ecosystems on private lands in the US South: an ecological economic analysis. *Ecological Economics*, 40, 411–419.

Anon (2009) http://www.ifer.cz/download/Infapro_carbon_inventory_en.pdf (accessed 20 September 2009.

Appanah, S and Turnbull, J.M. (eds) (1998) *A Review of Dipterocarps: Taxonomy, Ecology and Silviculture*, Centre for International Forestry Research, Bogor, Indonesia.

Asher, J., Warren, R., Fox, R. *et al.* (2002) *The Millennium Atlas of Butterflies in Britain and Ireland*, Oxford University Press.

Boyer, W.D. (1963) Development of Longleaf Seedlings Under Parent Trees. Res. Pap. SO-4. New Orleans; U.S. Department of Agriculture Forest Service, Southern Forest Experiment Station. Cited in Brockway, D.G. *et al.* (2005).

Brockway, D.G., Outcalt, K.W., Tomczak, D.J. and Johnson, E.E. (2005) Restoration of Longleaf Pine Ecosystems. General Technical Report SRS-83, USDA Southern Research Station. http://www.srs.fs.usda.gov/pubs/gtr/gtr_srs083.pdf (accessed 4 September 2009).

Chan, C.L., Lamb, A., Shim, P.S. and Wood, J.J. (1994) *Orchids of Borneo*, The Sabah Society and Royal Botanic Gardens Kew, Kota Kinabalu.

Chan, H.T., Shamsudin, I. and Ismail, P. (eds) (2008) *An In Depth Look at Enrichment Planting*, Forest Research Institute, Kuala Lumpur.

Contreras-Hermosilla, A. (1999) Towards Sustainable Forest Management: An Examination of the Technical, Economic and Institutional Feasibility of Improving Management of the Global Forest Estate. ftp://ftp.fao.org/docrep/fao/003/X4107E/X4107E00.pdf (accessed 20 September 2009).

Ellwood, D.F. and Foster, W.A. (2004) Doubling the estimate of invertebrate biomass in a rainforest canopy. *Nature*, 429, 549–551.

Falck, J., David, A., Garcia, C. and Yap, S.W. (2007) Rehabilitation of a degraded tropical rain forest in Sabah, North Borneo. Proceedings of Session 072: Harmonising Commercial Utilisation, Social and Conservation Values through Intensive Tropical Forest Management.

FAO (2009a) ftp://ftp.fao.org/docrep/fao/011/i0350e/i0350e.pdf (accessed 20 September 2009).

FAO (2009b) http://faostat.fao.org/site/630/default.aspx (accessed 20 September 2009)

Forestry Commission (2003) National Inventory of Woodlands and Trees. http://www.forestry.gov.uk/pdf/nigreatbritain.pdf/$FILE/nigreatbritain.pdf (accessed 20 September 2009).

Forestry Commission (2006) Forestry Statistics 2006. Available at www.forestry.gov.uk/website/ForestStats2006.nsf/byunique/imports.html (accessed 13 March 2010).

Garbutt, N. and Prrudente, C. (2006) *Wild Borneo. The wildlife and scenery of Sabah, Sarawak, Brunei and Kalimantan*, New Holland, London.

Hart, C. (1991) *Practical Forestry for the Agent and Surveyor*, 3rd edn, The Bath Press, Avon.

Hill, D.S. and Abang, F. (2005) *The Insects of Borneo*, University of Malaysia Sarawak, Kota Samarahan.

Hill-Cottingham, P., Briggs, D., Brunning, R. *et al.* (2006) *The Somerset Wetlands. An Ever Changing Environment*, Somerset Books, Tiverton.

ITTO (2002) ITTO Guideline for the Restoration, Management and Rehabilitation of Degraded and Secondary Tropical Rainforest. http://www.itto.int/en/policypapers_guidelines/ (accessed 24 August 2009).

ITTO (2005) Status of Tropical Forest Management, Country Profile, Malaysia, pp. 157–163. http://www.itto.int/en/policypapers_guidelines/ (accessed 24 August 2009).

ITTO (2007) Making SFM Work. ITTO'S First Twenty Years. http://www.itto.int/ (accessed 20 September 2009).

ITTO (2009) www.itto.int/en/feature04/ (accessed 20 September 2009).

JNCC (2009) http://www.jncc.gov.uk/page-5 (accessed 20 September 2009).

Krishnapillay, B. and Tompsett, P.B. (1998) Seed handling, in *A Review of Dipterocarps. Taxonomy, Ecology and Silviculture* (eds S. Appanah and J.M. Turnbull), Centre for International Forestry Research, Bogor, Indonesia. Pp. 74–88.

Lander, J.L., Van, L., David, H. and Boyer, W.D. (1995). The longleaf pine forest of the southeast: requiem or renaissance? *Journal of Forestry*, 93, 39–43.

Lee, Y.F. (2003) *Preferred Checklist of Sabah Trees*, 3rd edn, Natural History Publications, Kota Kinabalu.

Luyssaert, S., Schulze, E.D., Börner, A. *et al.* (2008) Old growth forests as carbon sinks. *Nature*, 455, 213–215.

Nakamura, K. (2006) Micropropagation of *Shorea roxburghii* and *Gmelina arborea* by shoot-apex culture, in *Plantation Technology in Tropical Forest Science* (eds K. Suzuki, K. Ishii, S. Sakurai and S. Sasaki), Springer, Tokyo.

National Geographic (2001) Loss of Amazon Rain Forest May Come Sooner than Expected. National Geographic News. Available at. http://news.nationalgeographic.com/news/2001/06/0626_amazonrainforest.html (accessed 15 March 2010).

NBN Gateway (2009) Lower plants species list. http://data.nbn.org.uk/ (accessed 20 September 2009).

O' Malley, K. (2009) Patterns of abundance and diversity in epiphytic orchids on Para-shorea malaanonan trees in Danum Vallyey, Sabah. Undergraduate thesis. Available at http://www.theplymouthstudentscientist.org.uk/index.php/pss/article/view/78/128 (accessed 15 March 2010).

Outcalt, K.W. (1998) Needs and Opportunities for Longleaf Pine Ecosystem Restoration in Florida, in Ecological Restoration and Regional Conservation Strategies (ed. J.S. Kush), Solon Dixon Forestry Education Centre, Andalusia. Longleaf Alliance Rep. 3.

Pahari, K. and Murai, S. (1997) Simulation of Forest Cover Map for 2025 and Beyond Using Remote Sensing and GIS. http://www.gisdevelopment.net/aars/acrs/1997/ts12/ts12003pf.htm (accessed 20 September 2009).

Parliament UK (2008) A memorandum submitted by the Woodland Trust, http://www.publications.parliament.uk/pa/cm200708/cmselect/cmenvaud/76/76we02.htm (accessed 20 September 2009).

Peet, R.K. and Allard, D.J. (1993) Longleaf pine-dominated vegetation of the southern Atlantic and eastern Gulf region, USA, in Proceedings of the 18th Tall Timbers Fire Ecology Conference, Tall Timbers Research Station, Tallahassee (ed. S.M. Herman). Cited in Brockway, D.G. *et al.* (2005).

Rackham, O. (1988) Trees and woodland in a crowded landscape: The cultural landscape of the British Isles, in *The Cultural Landscape; Past, Present and Future* (eds H.H. Birks, H.J.B. Birks, P.E. Kaland and D. Moe), Cambridge University Press, Cambridge.

Rackham, O. (2006) *Woodlands*, Collins, London.

Read, H.J. and Frater, M. (1999) *Woodland Habitats (Habitat Guides)*, Routledge, Abingdon.

Rodwell, J.S. (1998) *British Plant Communities*, Woodlandand Scrub, Vol. 1. Cambridge University Press.

Rose, F. (1981) *The Wildflower Key*, Warne, N.W. Europe.

Shukor, N. (2001) Status of *in situ* conservation of commercial tree species in Malaysia, in *Proceedings of the International Conference on ex situ and in situ Conservation of Commercial*

Tropical Trees, (eds B.A. Thielges, S.D. Sastrapradja and A. Rimbawanto), ITTO, Yogyakarta, Indonesia.

Smith, S., Gilbert, J. and Coppock, R. (2000) Great Britain: New Forecast of Softwood Availability. Forestry Commission Report. http://www.forestry.gov.uk/pdf/publishedforecast2000.pdf/$FILE/publishedforecast2000.pdf (accessed 06 August2009).

Soepadmo, E., Saw, L.G. and Chung, R.C.K. (2004) *Tree Flora of Sabah and Sarawak*, vol. 5, Sabah Forestry Dpt Malaysia, Forest Research Institute and Sarawak Forestry Dpt Malaysia, Kuala Lumpur.

Sunart Oakwoods Initiative (2009) www.sunartoakwoods.org.uk (accessed 18 July 2009).

Suzuki, K., Ishii, K., Sakurai, S. and Sasaki, S. (eds) (2006) *Plantation Technology in Tropical Forest Science*, Springer, Tokyo.

Syme, N. and Currie, F. (2005) Woodland Management for Birds; a Guide to Managing for Declining Birds in England, RSPB Sandy and Forestry Commission, Cambridge.

Tabor, R. (2000) *Traditional Woodland Crafts*, Batsford, London.

The Longleaf Alliance (2009) http://www.longleafalliance.org/ (accessed 20 September 2009).

Tompsett, P.B. (1998) Seed physiology, in *A Review of Dipterocarps. Taxonomy, Ecology and Silviculture* (eds S. Appanah and J.M. Turnbull), Centre for International Forestry Research, Bogor, Indonesia.

Vera, F.W.M. (2000) *Grazing Ecology and Forest History*, CABI Publishing.

Vince, S.W., Duryea, M.L., Macie, E.A. and Hermansen, L.A. (eds) (2005) *Forests at the Wildland-Urban Interface Conservation and Management*, CRC Press, Boca Raton.

Wan Mohd Shukri, W.A., Shamsudin, I., Samsudin, M. *et al.* (2008) Concept of enrichment planting in tropical forest, in *An In Depth Look at Enrichment Planting* (eds H.T. Chan, I. Shamsudin and P. Ismail), Forest Research Institute, Kuala Lumpur. Pp. 9–29.

Willis, K.L. and McElwain, J.C. (2002) *Evolution of Plants*, Oxford University Press.

Yahya, A.Z., Ghani, A.A., Koter, R. *et al.* (2006) *Establishment and Management of Khaya Ivorensis Plantation*, FRIM Technical Information Handbook No. 37, FRIM, Kuala Lumpur.

5 Farming and the aquatic environment

5.1 Water

Water is one of the simplest compounds known to man – two hydrogen atoms and one oxygen atom – but it is also probably the most important. Approximately three quarters of the planets' surface is covered by the oceans, and the land on which we live is interspersed by a network of freshwater and brackish watercourses and lakes. Our need for this resource is paramount and we exploit it indiscriminately, without foresight, adopting an attitude of 'Out of sight, out of mind'.

5.2 Water framework directive

Ninety percent of pollution of the coastal fringe occurs from the land. The EU Water Framework Directive (WFD) was established and implemented in December 2000. It is a comprehensive European directive that lays out plans to improve water quality in our waterways, water bodies, estuaries and underground water across the whole of the European Community. The requirement is for our rivers, lakes, ground and coastal waters to reach good ecological status by 2015 with a review and updates every six years thereafter. Implementation of the directive is to be achieved by adherence to a specific timetable and in a variety of ways.

The reduction of water pollution is clearly the main thrust of the required strategies and, although there are many sources of these pollutants, farming is one of the main industries that needs to be targeted to achieve full implementation of the directive. In conjunction with the WFD, the Common Agricultural Policy (CAP)[1] introduced incentive schemes for farmers including, single farm payments (SFPs) where subsidy payments are linked to cross compliance, agri-environment schemes to provide protection against diffuse pollution into waterways and higher-level schemes for more costly projects such as the creation of wetlands. Many areas of the UK are designated Nitrate Vulnerable Zones (NVZs), particularly those close to estuaries. Farmers are obliged as part of their cross compliance to adhere to further legislation

[1] NB Details of the WFD and CAP policies can be found on the DEFRA and Environment Agency (EA) websites.

Introduction to Wildlife Conservation in Farming   Edited by Stephen Burchett and Sarah Burchett
© 2011 John Wiley & Sons, Ltd

to protect their environment. New updates of NVZs apply as of January 2010; this is described in greater detail in Chapter 1.

Currently a scheme called River Basin Management Planning is being implemented whereby a specific plan for each of the 96 designated areas across Europe has an explicit plan to set out the environmental pressures and to identify the measures required to meet the WFD objectives for the rivers, lakes, estuaries, coastal waters and underground water within that area.

In this chapter we look at the importance of waterways in the farming environment and to the environment as a whole with respect to obligate aquatic wildlife and other wildlife that is dependent on an aquatic phase.

We will also look at aquaculture in detail with respect to its environmental impact; its role in food security and what is currently being done to improve the methods being used.

- Part 1: On farm ponds, watercourses and riparian strips
- Part 2: Fens, marshes and wetlands
- Part 3: Estuaries, Coastal and Marine
- Part 4: Aquaculture.

5.3   Part 1: On farm ponds, watercourses and riparian strips

5.3.1   Water contamination from soil erosion and agricultural runoff

Regardless of the size and farming system employed, soil is a fundamental resource for any farmer and should be treated carefully and with respect. Sustainable soil management is the key to successful and profitable yields and good soil structure helps to maintain crop health and reduce the risk of flooding.

The development of modern agriculture, with its large-scale heavy machinery, grubbed-up hedges and prophylactic use of pesticides, and nutrient applications, has led to a subsequent decline in soil structure and fertility through deterioration in soil organic matter and run-off. As a consequence there has been an increase in the incidence of major soil erosion events in the winter and spring months. For example, in 2000, the UK experienced one of the wettest autumns on record that coincided with the late planting of winter crops, thus the soil surface was left exposed and vulnerable to soil erosion. Extensive soil erosion was observed throughout the British Isles, in one area of note in Steppingley, Bedfordshire, an area of approximately 4 ha was extensively rilled and gullied with estimates of 10 tonnes of soil lost from one gully alone (www.cranfield.ac.uk/sas/nsri – accessed May 2009).

Annual soil erosion rates in many areas of the world are extremely variable but on average are estimated to be in the region of 17 tonnes/ha/year. That greatly exceeds the rate of soil formation (1 tonne/ha/year) and research has demonstrated that

2.3 million tonnes per year of topsoil are lost across the UK with 44% of arable land being particularly vulnerable. Soil erosion rates in the US are thought to average 10 times the replenishment rate and in parts of India and China it is thought to be closer to 30–40 times faster than the replenishment rate (Pimentel, 2006).

These rates of soil loss are clearly unsustainable and have motivated policy makers into action. The first draft of a Soil Action Plan for England: 2004–2006 was published by DEFRA such that elements of the plan featured under cross compliance and the current agri-environment schemes. The plan has more recently been updated and lays out a four part strategy: The sustainable use of agricultural soils; the role of soil in mitigating and adapting to climate change; protecting soil function during construction and development; and preventing pollution and dealing with our legacy of contaminated land (DEFRA, 2009).

The reform of CAP from production-based subsidies to the SFP places a greater burden on farmers and growers to implement codes of good agricultural practice including elements of soil protection under cross compliance.

5.3.2 Origins of soil erosion

There are two main causes of soil erosion: water and wind. Water erosion is the most significant cause of soil erosion in the UK and it is the result of heavy winter rains on fields at field capacity and, in particular, where soils have poor structure. This would apply in general to other areas around the world where rainfall is typically high. Poor soil structure reduces the drainage rates of surface water and results in surface water run-off carrying and depositing soil particles into water courses (Figure 5.1) and onto roads. The second type of soil erosion is caused by wind and occurs mainly on light exposed chalky soils in the spring.

Current agricultural practice has resulted in a significant increase in the degradation of soils, in particular soil structure and in the degree of soil organic matter.

Figure 5.1 Brown colouration in a stream resulting from soil erosion. (Photo: Stephen Burchett)

The following farming practices can lead to the destruction of soil structure and consequently an increased risk of soil erosion:

- late sown winter cereal crops
- row crops such as potatoes, onions and maize
- vegetable crops such as cauliflower, especially when left as a stubble
- high stocking rates in wet areas
- feeding and drinking troughs and
- supplementary feeding.

5.3.3 Economic and environmental implications of soil erosion and agricultural runoff

The financial implications of soil erosion are significant and include the loss of:

- valuable nutrient rich top soil
- reduced rooting volume
- poor crop performance
- the requirement for extra cultivations to repair the damage
- prosecution by the authorities for environmental pollution.

The cost to local habitats and wildlife is also significant and include:

- deposition of sediment onto roads
- damage to watercourses from inputs of agro-chemicals
- sediment in rivers that damage fish spawning grounds
- greater flood hazard.

5.3.4 Influence of soil structure on soil erosion

Soil structure is a measure of the arrangement of soil particles into crumbs and blocks that make up a soil matrix. The growth of plants further modifies this structure by root expansion that opens up soil pores and leads to additional diversity in soil particle size. The action of weather on soil structure also modifies the nature of soil via wetting, drying, freezing and thawing.

The texture of soil is a physical measure of the building blocks of soil which range from sand (0.2–2 mm), silt (0.002–0.06 mm), to pure clay (<0.002 mm) with variations on the basic texture, such as a clay-loam. The structure of soil is thus

Figure 5.2   An example of good soil structure. (Photo: Stephen Burchett)

influenced by the textural classification, if a soil is classified as clay the structures are more tightly packed and thus less likely to break apart. Clay is notoriously difficult to work and compacts if worked when wet, and cracks in dry summers. On the other hand sandy soils are easy to work, warm up quickly in the spring and are freely draining, but they slump and are easily eroded via water and wind.

Good soil structure (Figure 5.2) is important because it provides the correct conditions for plant roots and thus has a direct influence on crop growth and performance. A well-structured soil will:

- aid water and air flow
- increase soil fauna
- improve germination and establishment
- reduce soil erosion.

Poor soil structure (Figure 5.3) causes increased risk of:

- soil erosion
- poor germination and establishment
- nutrient leaching
- increased run-off of agricultural chemicals
- increased risk of environmental pollution.

Figure 5.3   (a,b) Poor soil structure – soil is compacted. (Photo: Stephen Burchett)

5.3.5   Impacts of soil erosion and agricultural runoff on aquatic wildlife

It is estimated that approximately 60% of soil that is washed away ends up in waterways and ponds therefore, as well as having detrimental effects on crop growth and thus on the economy of the farm, soil erosion and agricultural runoff can devastate aquatic life. This can occur in a number of ways.

- Sedimentation
- Poisoning
- Eutrophication.

Sedimentation

This can silt up the watercourse decreasing volume, depth and thus affecting flow dynamics. This does not only adversely affect the drinking water for stock, but can also change the environmental conditions for the resident wildlife. For example, in Europe the caddis fly (*Trichoptera spp.*), spends its larval stage in a case that it makes from tiny stones, sand and shells on the streambed. Though the caddis fly larva can move around to search for food it mostly relies on the flow of water to carry its food along in the current.

Another victim is the freshwater crab. Of the 1280 known species, around 16% are considered endangered, although due to the current lack of data this could be a much higher figure. Most fresh water crabs live in the tropics, mainly in SE Asia. Increased human populations have engendered an increase in habitat loss due to increased development and agriculture. Sedimentation of freshwater sites resulting from these activities is becoming an increasing problem as a result, and the crabs are suffering. A BBC article (July 2009), highlighted the plight of a crab that is indigenous to Singapore, the *Johura singaporensis*, which is particularly vulnerable to such disturbances as it lives on an island (Walker, 2009).

Also, in the tropics, sedimentation is a cause of major concern where mangroves have been felled for timber harvesting and the charcoal industry to Japan from Malaysia (www.mangroveactionproject.org – accessed June 2009). Mangrove trees bind the soil in coastal areas, so that where they are felled, great swathes of soil sediment is washed directly into the sea, swamping vulnerable sea grass beds and coral reefs. The resulting damage to the corals is well documented, but it should also be noted that both the mangrove swamps and the sea grass beds function as nurseries for reef species. These trees are not ostensibly being 'farmed', only harvested, but here is an opportunity for farming in the form of forestry to play an important role by replanting. Although the Malaysian government has a replanting programme, it by no means replaces the trees being felled (www.mangroveactionproject.org – accessed June 2009). Mangroves constitute a unique habitat both above and below the tide line, no other harvestable crop will grow in this habitat so this is a perfect chance for farming and wildlife conservation to function in harmony.

However, it should be noted that the right answer for one situation, might be quite wrong for another. The Kammu people of Laos traditionally consider that reforestation on the swidden lands that they farm would cause streams to dry up, harmful to them and to the aquatic life of the stream, but also potentially causing water levels downstream to be affected (Rambo, 2007). Each situation must be regarded on its own merits.

Poisoning

Chemical pesticides, herbicides and concentrated runoff from manure can devastate wildlife if any of them find their way into watercourses. For example, despite clear evidence of the detrimental effects on human health and its possible link to cancer, the herbicide Atrazine, which has been banned in Europe since 2003, is still one of the most commonly used herbicides in the world. It has been linked to declining aquatic reptile populations for many years as it is a known endocrine-disrupting compound (EDC), and a new study by biologists at the University of California, Berkeley, US, using African Clawed Frogs (*Xenopus laevis*) shows that it directly causes a sex change from male to female. Many other similar examples from around the world are well documented.

Most farmers, apart from organic farmers, apply chemical pesticides and herbicides to their fields to control deleterious insects, fungi and to control weed incursion. These chemicals can be expensive to buy and to apply; it is therefore in the economic interest of the farmer to use no more *quantity* than is required, and to apply no more *often* than is needed and to take note of any advice on how applications should be made. However, the use of these chemicals is not necessarily well managed in all cases. It is highly recommended that applications are made on a still day. Where this is done on a windy day, spray drifts in the wind, not only reducing the amount being applied to the crop, but damaging the surrounding environment. Field margins and hedges are commonly damaged in this way, but the spray drift can also find its way into watercourses poisoning plants and animals. Where such chemicals are applied in excess, more will accumulate in the groundwater, and in surface soils vulnerable to soil erosion.[2] Runoff from ill-managed manure from stock will have similar effects on wildlife associated with these aquatic environments.

An extreme case of pollution from manure runoff occurred in Beaver Creek, Minnesota, US in 1997.[3] 100 000 American gallons ($\sim 3.8 \times 10^5$ l) of manure from an intensive pig farm washed into the creek. The immediate impact was devastating, with an estimated 690 000 fish deaths alone and no possible way to record the invertebrate fatalities. But it didn't end there. Beaver Creek runs into the Minnesota River and then into the Mississippi. Two days after the spill and many miles away a dairy farm whose cows drank from the river water had to destroy all its milk, and the pregnant cows aborted their calves. Coliform levels in the river alongside this dairy farm were still 3800× that which is considered the maximum for safe drinking water (Jackson and Jackson, 2002).

---

[2] In the UK the Government has an excellent website illustrating what the regulations are with respect to diffuse pollution and ways to prevent it. See www.netregs.gov.uk.

[3] NB The United States has a policy called the Clean Water Act; this was first established in 1977, though sub-sections of the act are regularly modified. Details can be found on the US Environmental Protection Agency website.

Eutrophication

This is a process that occurs where excessive nutrients, mainly Nitrogen (N) and Phosphorous (P) accumulate in still water such as ponds and still areas on the fringes of slower moving watercourses. Although eutrophication can occur naturally, it is caused mainly by run-off from poorly managed farmland and from anthropogenic amenities such as sports grounds that are often fed excessively to maintain the grass. Because these nutrients are not always easy to trace back to their source as they result from seepage, runoff and flow along the watercourse, these are known as non-point source (NPS) pollutants.

As water plants vary in their capacity to compete for these nutrients, rapidly growing algae on the surface of the water proliferates, as does heterotrophic fungi and bacteria, blocking out sunlight to the other plants in the water column and bed below. This becomes a further problem when the algae dies off and decays reducing dissolved oxygen (DO) in the water – that is the water becomes hypoxic. This system becomes self-perpetuating because, though the algae produce oxygen as a by-product of photosynthesis, most of this will be lost to the atmosphere as oxygen gas, and at night, the respiring algae consume any available DO. As the water is still, there is little diffuse atmospheric oxygen to be had. Some aquatic invertebrates such as bloodworms (*Glycera spp.*) and fish such as those in the *Cyprinid* family, for example the roach (*Rutilus rutilus*), are more tolerant than others of low DO levels, subsequently significant changes in community structures and substantially decreased biodiversity can result. A further consequence is that as a pond becomes more and more silted up, encroachment and succession of land plants may ensue culminating in the pond disappearing altogether (Figure 5.4).

Later in this chapter there is a case study looking at a modern salmon farm in Scotland. One impetus for the increase in salmon farming has been the devastating decline in wild salmon stocks. Salmonids are extremely intolerant of watercourses with low levels of DO, so that where their migratory routes take them through such waters many, if not all, will not survive to reach their breeding grounds. This, coupled with the damaging effects of lice infection from many of the more intensive salmon farms, and the use of hydroelectric systems in many such rivers, has resulted in a substantial decline in wild salmon stocks, estimated to be in the region of an 80% decline, as well as the decline of other salmonids such as sea trout (*Salmo trutta*) (Rawlings, 2006).

5.3.6 Management of soil erosion and agricultural runoff

There are several strategies for managing soil erosion and these should be employed as an holistic approach to integrating soil and waste management with conservation of the resident wildlife and the local environment. This starts with the construction

Figure 5.4 Eutrophied pond, surrounding vegetation is encroaching and needs clearing to maintain. (Photo: Stephen Burchett)

of a soil erosion risk map. The following strategies can be employed across a holding to improve soil structure and reduce the risk of soil erosion:

- Develop stable topsoils by increasing soil organic matter
    — Farm Yard Manure (FYM) must not exceed 250 kg/ha/year in Europe on conventional farms and 170 kg/ha/year on organic farms
    — If the farm is on a NVZ then organic nitrogen must not exceed 250 kg/ha/year in grass crops and 210 kg/ha/year in non grass crops falling to 170 kg/ha/year after four years in the NVZ
    — Account for nutrients in FYM and crop available nitrogen when applying inorganic fertilisers
- Conservation tillage systems (discussed in Chapter 2 above)
- Avoid surface compaction
    — Keep off fields when soil is at field capacity
    — Correct compaction before sowing
    — Subsoil

- Protect soil during autumn and winter
- Timely sowing of autumn crops
- Cover crops
- Chisel plough to leave a rough and uneven surface, which aids percolation
- Create buffer zones, bunds and sediment traps
- Manage riparian strips in favour of local wildlife.

5.3.7   Management of riparian strips

We visited a mixed organic farm in Peel Forest in the South Canterbury region of South Island, New Zealand. The land had a creek running the length of the farm, here the farmer was able to obtain some financial help via covenants associated with the Queen Elizabeth Second Resource Management Initiative, and received a one-off payment to plant riparian strips. This is particularly pertinent in New Zealand where much of the land surface is dominated by introduced species, however, many native species of fish and invertebrates live on in the rivers, creeks and streams of New Zealand. Native freshwater biodiversity is low (www.nzfreshwater.org – accessed June 2009); therefore maintaining what does exist is a significant conservation issue. This creek and the flanking strips were healthy and teeming with invertebrate life, *Dipteran* and Dragonfly (*Odonata*) species were particularly noticeable and the farmer reported the presence of the 'koura', the local name for the native crayfish (*Paranephrops planifrons*) and the native longfin eel (*Anguilla dieffenbachia*).

A visit to Arbigland Farm Estate in Dumfriesshire, SW Scotland also revealed the management of riparian strips. Here a 'burn' (Scottish term for a stream) runs along the boundary. A 12 m riparian strip borders the waterway and is managed for wildlife in compliance with the regulations associated with farming on a NVZ. To one side of the burn, one of the fields lays wet for around six weeks a year, encouraging the presence of numerous birds, otters and hares. This field is managed by grazing in the summer when it dries out. In these burnside areas, reed buntings (*Emberiza schoeniclus*), can commonly be seen, as well as curlews (*Numenius arquata*) (Species Box 5.1), lapwings (*Vanellus vanellus*) and meadow pipits (*Anthus pratensis*). Bat numbers have increased as a result of management strategies and it is noticeable that invertebrate numbers have also increased, although no surveys have yet been carried out to quantify this observation. This farm is discussed in greater detail later in the text.

Restoration of wetland areas is an important step to increasing biodiversity on farmland. We visited an inspiring project in North Carolina, US. The Kuenzler Farm Wetland Reserve is the subject of Case Study 5.1 below.

Species Box 5.1: Curlew (*Numenius arquata*)

Profile:

Image: Curlew © Sandra Hughes 2009

Wading bird of the *Scolopacidae* family, it is the largest wader in Europe at around 50–57 cm in length with a wingspan of 1 m, and both sexes and the juveniles look alike. In the UK it can be seen all around the coastline, and it also ranges right across the temperate regions of Europe and Asia. Its name comes from its call, and it is recognisable from its long down-curved bill, which is longer than that of its relation, the Wimbrell. The curlew eats mainly marine worms and invertebrates including small crustacea, though it moves further inland, usually onto moors, meadows, marshes and sand dunes, where earthworms become a major part of their diet, to breed in the spring and early summer. The female lays four eggs, the young leave the nest with the parents shortly after hatching and they can fly after five to six weeks.

Conservation Status:

- RSPB (Royal Society for the Protection of Birds) Amber status
- Red List for Birds of Conservation Concern in Ireland
- Annex 1 EC Birds Directive
- Appendix 2 Bern Convention
- International Union for Conservation of Nature (IUCN) – Near-threatened status 2008.

Current Status:

There has been a general decline in curlew numbers due to loss of habitat to agriculture. This decline has been more serious in Northern Ireland where numbers are thought to have decreased by more than 50% in the last 25 years.

References

Peterson, R.T., Mountfort, G. and Hollom, P.A.D. (1983) *A Field Guide to the Birds of Britain and Europe*, 4th edn, William Collins & Co Ltd, Glasgow.

www.rspb.org.uk

www.peatlandsni.gov.uk

Case Study 5.1: Kuenzler Farm Wetland Restoration Programme

We won't be the first people to visit, or indeed to write up a case study about an inspiring farm in Orange County, North Carolina, US, and we certainly won't be the last. We were accompanied to the farm by Brent Bogue of the US Department of Agriculture (USDA) and John-Ann Shearer of the US Fish and Wildlife Service (FWS). Once there we found Jutta Kuenzler keen to tell her story about the restoration of 17 acres (6.8 ha) of wetland on her 154 acres (62 ha) farm. As such it has been visited by politicians, press and innumerable school children, all eager to learn about the benefits of introducing native species and restoring the habitat to support them.

Ed and Jutta Kuenzler bought their farm in the mid 1960s. It is a mixed farm with rich hay meadows, pastures and forest, they also grow corn and beans and raise beef cattle. Much of the farm had been planted with fast growing grasses such as fescues (*Festuca spp.*) and with loblolly pines (*Pinus taeda*) for timber.

Ed was an Environmental Sciences Professor at the University of North Carolina and often used his farmland for field trips for his students. In the early days the land was abundant with native wildlife such as the wild turkey (*Meleagris gallopavo*), the Whip-Poor-Will (*Caprimulgus vociferous*), the coyote (*Canis latrans*) and the Bob White Quail (*Colinus virginianus*) (Species Box 5.2). Over time, a decline in such species was becoming apparent and Ed worked on identifying all the species of plants and animals to be found on their land, recording numbers and movements using a 6-hour observation pattern, and making seasonal observations too, as migrant birds are of particular abundance in the Carolina's. Ed recorded over 400 species, clearly this was a land worth preserving for the future, and a seed of any idea began to grow. A 17 acres (6.8 ha) parcel of land on the farm near Collins Creek had previously been wetland and had, at some point in the past, been drained and ditches dug so that it could be used as pasture. The Kuenzler family decided that this land should be restored to its original state in order to encourage wetland birds, amphibians, reptiles and native wetland flora.

With the help of the Natural Resource Conservation Service (NRCS), FWS and the Orange County Soil and Water Conservation District, a long and involved process began. Seeds were collected from the soil seed bank and grown up by a local nursery (Niche Gardens Nursery), the FWS provided $4000.00 for this part of the project. Then the area was scraped and the ditches plugged. No one knew exactly how the land would have looked before it was drained, so the team devised a plan to flood the area. Low areas were scalloped out alongside the existing ditch, three gently shelving shallow pools were excavated, water flow was diverted to fill the area via these pools and then planting could take place to complete the mission using plants grown by the nursery, both those that had been collected previously, and many other native plants (the nursery's speciality). Natural colonisation took place over time too. The wetland area was enrolled into a permanent Wetlands Reserve Program (WRP) easement by the NRCS who then funded the rest of the restoration costs of $13 000.00. Sadly Ed Kuenzler died in 2001 while the work was still ongoing, but Jutta Kuenzler and her daughter Doreen continued with the project and completed it in 2002.

Species Box 5.2: Bob White Quail (*Colinus virginianus*)

Profile:

The Bob White is named so because of its unusual call. It is a ground-dwelling game bird with a range that extends across the US, Mexico and the Caribbean. It weighs between 140 and 170 g and measures some 200–250 mm in length. For most of the year the Bob Whites live in groups called 'coveys' comprising some 10–30 birds, although during the mating season they go off in pairs. Nesting in tall grassland, the female lays 10–15 eggs in a clutch, one laid per day, these are heavily predated, and the pair may have three or four clutches a year if the eggs in the early clutches do not successfully hatch. Hatching occurs after 23 days, and the chicks are independent after only one day. Their preferred food is seeds,

Photo: Bob-White Quail © Don Brockmeier

especially grasses and cereals, but they also eat fruit, leafy vegetation and invertebrates. They, in turn, are preyed on by coyotes, foxes, raccoons and birds of prey, all of which will take the birds and their eggs. They are however very well camouflaged in amongst the tall grasses. Away from the nesting season they are also found living amongst scrubby, early successional growth, where young shoots and fruits are close to the ground. They are often found close to waterways or ponds, though they can gain much of their requirements for water from fruit and dew.

Conservation Status:

NA

Current Status:

Wild populations are under some degree of threat from the loss of habitat, although their popularity both for game hunting and their eggs, which are considered a delicacy, ensures their success due to high density captive breeding.

As planned, the wetland area requires very little management input, and is left largely to its own devices. This has yielded tremendous rewards; the sound of frogs calling accompanied our walk around the wetland area. There were a number of species there including the Northern Cricket Frog (*Acris crepitans*) (Species Box 5.3).

Many amphibians, as well as invertebrates, fish, birds, reptiles and mammals now occupy the site. With respect to the national campaign in the US to increase plantings of the long leaf pine (LLP) (see Chapter 4), whilst decreasing the more persistent, weedy loblolly pine that occupies much the same niche, we made an interesting observation during our visit

Species Box 5.3: Northern Cricket Frog (*Acris crepitans*)

Profile:

A tiny frog, between 16 and 35 mm in length, the Northern Cricket Frog belongs to the tree frog family (*Hylidae*), though it is not itself arboreal. Consisting of three sub-species, the frogs are widely distributed, ranging across most of North America and NE Mexico. They feed on small invertebrates and are preyed on themselves by fish, birds and small mammals. They can leap several feet in order to

Photo: Northern Cricket Frog © Stephen Burchett 2008

avoid predators and can be seen to zigzag out of harm's way. They mature in their first year and breed between May and July, laying their eggs individually on vegetation. They prefer fairly shallow water, are efficient swimmers and can be found in both permanent ponds and slow moving waterways.

Conservation Status:

IUCN Red List – Threatened – 2008.

Current Status:

In decline throughout their range, although locally abundant in Orange County NC, US. They are an indicator of good water quality being intolerant of chemical inputs, which have certainly contributed to their decline. They are also prone to fungal disease and to drought.

to the farm. Beavers (*Castor Canadensis*) are starting to move back into the area. There exist polarised views on beaver colonisation. Considered a pest by some landowners, and a desirable addition to the suite of wildlife to be seen by other landowners, the beaver none the less belongs in this environment. Jutta Kuenzler was delighted when our visit revealed the presence of beavers on her land, this wasn't difficult to spot as there was a beaver dam under construction! It was noted that the beaver or beavers were gnawing the loblolly saplings (Figure 5.5), and leaving the LLP saplings alone. We have no idea whether this is a recognised phenomenon, or even if it has ever been observed before or since, certainly we found no such data. However, if this behaviour persists, the long leafs seem to have a future in this part of North Carolina, it may even be one of the factors that made the long leaf more prevalent in the past. Perhaps it is the nature of such observations to present more questions than answers!

The Kuenzler family have now put all of their land into permanent conservation easement with both the Triangle Land Conservancy (TLC) and WRP. This means that this land is preserved forever, while it can be sold, it cannot be subdivided or developed.

Figure 5.5   Loblolly sapling that has been gnawed by a beaver. (Photo: Stephen Burchett)

5.3.8   Further downstream

These issues of soil erosion and agricultural runoff do not stop at fresh water. Water flow from farms often continues out to sea, sometimes directly, but usually via estuaries. Estuarine ecosystems are highly dynamic and complex and can be devastated by toxic algal blooms resulting from eutrophication, as can shallow coastal seas and marshland.

5.4   Part 2: Fens, marshes and wetlands

These areas represent a meeting point where terrestrial habitats meet aquatic habitats. They are complex, dynamic, marginal systems that are generally unique. Historically the value of these habitats was not normally recognised and such marginal land was often drained to increase pastureland for stock. One such area is the Fenlands of Eastern England.

5.4.1   Fenlands of Eastern England

Some of the most recognisable features of the Norfolk Broads, and parts of the surrounding counties, are the numerous windmills. Themselves, the subject of conservation initiatives due to their place in English industrial history and architectural archaeology, these constructions were used, not just for milling, but also for drainage on a landscape scale during the eighteenth and early nineteenth centuries, after which powered pumps began to take their place, and indeed, continue to do so today. When the land was drained, highly valued peat-based soils became available for agriculture and the area still remains one of the most productive in the UK. With a population in excess of 60 million and rising, and development high on the list of government priorities, productive land is at a premium and it becomes difficult to find a balance between agricultural productivity and conservation in such areas. However, there is yet an impetus to restore and maintain some areas of the fens for wildlife.

The Large Copper butterfly (*Lycaena dispar*) has gone from the area due to habitat loss. The native sub-species (*Ld dispar*) is sadly now extinct but work is currently being carried out to introduce the Dutch sub-species *Ld batavus* back into the fens (Asher *et al.*, 2001). Experiments carried out in the Weerribben National Park in the Netherlands, has indicated that summer cut fen meadow is favourable habitat for the Large Copper to breed successfully (Pullin, 1997). This gives hope that, not only can this butterfly be reintroduced successfully, but that fen grassland can continue to be harvested by the landowners. Again, farming and wildlife conservation can be seen to be mutually beneficial. Species Box 5.4 profiles this butterfly.

5.4.2   The Camargue

The Camargue is a vast delta wetland in the south of France covering an area of 145 300 ha, with an 80 km seaward edge bordering the Mediterranean Sea. Most of the Camargue delta falls between two arms of the Rhone and is flanked to either side by the Gulf of Aigues-Mortes and the Gulf of Fos. The northern-most areas are dominated by rivers, freshwater marshes and lakes with riparian woodlands, riverine dunes and some permanent grassland. Further lakes and rivers with some dry areas, brackish wet areas with variable salinity and some submerged brackish areas in the central areas; and saline lagoons to the south. Parts of the Camargue have been a nature reserve since 1927, the region attained National Park status in 1972 and it is now classed as a Biosphere Reserve (www.tourduvalat.org – accessed May 2009, www.lauguedoc-france.info – accessed June 2009).

Approximately a third of the Camargue is farmed, roughly 50 000 ha, the predominant crop being rice. There are also cereal crops, grapes, asparagus and other vegetables, reed harvesting, some arboriculture and pastureland for sheep, the Camargue beef cattle and the Camargue horse – the stock living semi-free roaming lives. Most farming in the region is relatively recent and initially quite damaging because of some wetland drainage and by the use of chemical inputs. Though still intensively farmed in many parts of the region, other areas have always been prohibitively difficult to turn to agriculture, plus conservation has become a key issue and the value of this unique habitat is now largely acknowledged both for its own sake and for the benefit of the tourist industry. The region boasts a recorded 75 fish species, 10 amphibians, 15 reptiles and nearly 400 birds as well as in excess of 1000 plant species (www.tourduvalat.org – accessed May 2009). Conservation groups have been

Species Box 5.4: Large Copper Butterfly (*Lycaena dispar*)

Profile:

Large copper-orange butterfly with silver trim and white to pale blue underwing in females. The males typically have fewer markings. They are heavily reliant on Water Docks (*Rumex hydro-lapathum*) on which the adults lay their eggs and the larvae feed. With a wingspan of up to 52 mm, they fly in August and are dependent on the heat of the sun to carry out their life cycle. They over-winter as second instar. Males are territorial.

Photo: Large Copper Butterfly © Peter Eeles 2005

Conservation Status:

(*Lycaena dispar*) three sub-species:

1. *L.d Batavus* – Biodiversity Action Plan (BAP) Species, Annex IV Bern Convention, Annexes II and IV EC Habitates and Species Directive – indigenous to the Netherlands.

2. *L.d Rutilus* – BAP Status – *Widespread throughout Europe, though in decline.*

3. *L. d Dispar* – indigenous to UK – extinct since the mid nineteenth century.

Current Status:

Reintroduction programme in progress for the Dutch sub-species *L D Batavus* to colonise the fens of Eastern England.

References

www.ukbap.org

www.ukbutterflies.co.uk

Pullin, A.S. (1997) *Journal of Insect Conservation*, 1 (3), 177–185.

working hard to protect the Camargue from further damage and to help restore habitats, and a number of projects are currently in hand. A good example can be seen with some work that has been carried out studying the European Bittern (*Botaurus stellaris*) breeding habits. Where in the past reed harvesting has been considered damaging to wetland habitats, observations indicate that bitterns have a preference for areas that have 10% open water enclosed by one-year reeds (Poulin *et al.*, 2009). This indicates that should the reeds not be regularly harvested this preference would not be realised. Thus the tradition of reed harvesting can be seen to be beneficial to the bittern (Species Box 5.5).

Species Box 5.5: European Bittern (*Botaurus stellaris*)

Profile:

A medium sized member of the heron family, the European Bittern is a very secretive bird rarely seen by all but the most determined. Seventy-five cm in length with broad wings spanning 130 cm and weighing in at around 1–1.5 kg. It has a wide range, from southern UK, across most of Europe and parts of northern Africa, Asia and central Russia. The males 'boom' in spring to attract mates, the booming can be heard a kilometre away, and individuals have a distinct 'voice', which can be recognised and monitored by conservationists. The bitterns live and nest in reed beds, mainly phragmites, but are highly vulnerable to pollution. Often active at night, they eat mainly fish, amphibians and insects.

Image: European Bittern © Sandra Hughes 2009

Conservation status:

- Bap Species
- Schedule 1, Wildlife & Countryside Act 1981
- RSPB Red List – Rare in UK, Globally of 'Least Concern'
- AEWA Action Plan Species category 3
- European Species Action Plan.

Current Status:

Globally less than 54 000 breeding pairs of which ~25 000 in Europe. Endangered due to loss of habitat, that is drainage of wetland areas for agriculture, water extraction for irrigation and abandonment of traditional uses of reed beds that hampers reestablishment of habitat.

NB. New book available: *The Bittern in Europe: A Guide to Species & Habitat Management*. 2006. This publication is the result of a cooperative project led by the Das Landesumweltamt Brandenburg (Brandenburg LUA), RSPB and Ligue pour la Protection des Oiseaux (LPO), and is available from the RSPB.

References

www.rspb.org.uk

African-Eurasian Waterbird Agreement (AEWA) website – www.unep-aewa.org

5.5   Part 3: Estuaries, coastal and marine

Estuaries are very complex and dynamic habitats. They provide the interface at the coast between the freshwater rivers or channels reaching the sea and the salt-water marine environment. It presents a gradient of fresh water, brackish water and marine water of varying depths, salinity and movement with every minute of every day dependent on tidal flow from the sea and river flow from inland. In turn, the freshwater source is dependent on high rainfall, particularly where rivers have numerous tributaries, or low rainfall, especially where extraction from water-courses is used for irrigation, and so on. Added to this are the numerous sources of potential pollutants into an estuary, land-based point source and NPS pollution from agriculture and other industries as well as ground water seepage and atmospheric pollutants, plus often high levels of traffic, all add to the mix. Inputs into estuaries, coastal and marine areas as well as fresh water sites are targeted by the WFD.

Working life can be very hard for farmers whose land lies close by estuaries. The soils are usually poor and shallow and are subjected to salt spray, and winds straight off the sea often unchecked by topographic obstacles, all of which can be detrimental to crops. This coupled with the pressures of adhering to the convoluted array of policies and legislation associated with farming in this environment makes farming here an extremely challenging undertaking.

When we visited Arbigland Farm Estate in Dumfriesshire SW Scotland, we met Jamie Blackett, the estate manager. Arbigland Estate is a mixed farm producing cereals, grassland for grazing and fodder crops in arable rotation. They also have sheep, a few horses and a pedigree herd of Luing beef cattle, there is also around 80 ha of mixed woodland, managed for the cattle and for pheasants and timber, there are a number of ponds on the estate. They have further diversified into holiday lets and the estate has been used as a prime film location. Arbigland even holds a place in American history. One of its gardeners in the eighteenth was father to John Paul Jones, founder of the US Navy. John Paul Jones was born on the estate.

The estate is bordered along its southern side by the sea, Jamie and the other people working on the estate not only manage the riparian strips, discussed above, but have to carry out their duties amidst a bewildering package of cross compliance regulations, plus recommendations from other bodies such as the RSPB. As a result, there are many conservation practices evident around the estate. Summer grazing is partially restricted; hedges are left to grow for longer before cutting and some of the crops, including a birdseed mix, are left unharvested to encourage wildlife. The RSPB carry out regular surveys at Arbigland and bird diversity is substantial, boasting a number of warblers, owls, martins and seabirds. Despite its maritime location, the estate is successful and productive. Although it is not an organic farm, inputs have been substantially reduced as they recycle nutrients efficiently by using farmyard manure and by growing legumes such as lupins in their crop rotation. It is a highly productive farm, where wildlife also thrives and, as well as the huge diversity of birds recorded by the RSPB, there have been many sightings of red squirrels (*Sciurus vulgaris*) in the woodland. However, perhaps one of the estate's best wildlife success stories is the presence of the Natterjack Toad (*Epidalea calmita*) rare in the UK; the estate is one of the few strongholds for this amphibian. Unusual amongst toads, the Natterjack is tolerant of brackish water; it is therefore ideally suited to the semi-maritime environment of Arbigland (Species Box 5.6).

Species Box 5.6: Natterjack Toad (*Epidalia calamita* – Formally *Bufo calamita*)

Profile:

Image: Natterjack Toad © Sandra Hughes 2009

The Natterjack Toad is one of only two toad species in Britain. Small, 7–8 cm, it is distinguished by a yellow stripe down its back. Nocturnal by nature, it burrows by day and forages by night, feeding on insects and other small invertebrates. The toad is, in turn, preyed upon by foxes, otters and large birds such as gulls and herons, though they produce a toxic secretion from their paratoid gland which offers a limited protection. Emerging from winter hibernation, the males have a loud, distinct boom to attract mates that can be heard several kilometres away. The females spawn in shallow ponds, as, unlike common toads, they are poor swimmers. Tadpoles develop quickly, often within a week, and the adult toad develops from six to eight weeks. Unusual for toads, the Natterjack is tolerant of brackish water and often inhabits dune areas. They live for 12–15 years.

Conservation status:

1. UK BAP Species
2. Protected under the Wildlife & Countryside Act 1981
3. IUCN Red List – category, Least Concern 2008 (though population in decline).

Current Status:

Though relatively widespread across Europe, the Natterjack Toad is suffering a general decline due to loss/changes in habitat. Very rare in Britain and therefore subject to conservation efforts, particularly in the Solway Firth in Scotland.

References

Authors various (1981) *AA Book of the British Countryside*, 2nd edn, Drive Publications, London.
www.bbc.co.uk/nature
www.snh.org.uk

The current implementation of the WFD includes a scheme called 'River Basin Management Planning', mentioned above. These plans were required in paper form by the end of 2009. Of the 96 districts identified across Europe as areas requiring such a plan, 14 are already up and running as pilot schemes. One of these is the Odense River Basin on the island of Fyn in Denmark. As well as directly measuring nutrient levels, depth of eelgrass (*Zostera marina*) growth is being used as an indicator species to assess the required reduction of pollutant inputs into the Odense Fjord. This recent field study, using historical data as a starting point, indicates that a reduction of 1200 tonnes N pa (11 kg N ha^{-1} in the catchment area), which is approximately 60% of this fundamentally wasted material into the fjord, is required to meet the minimum requirement laid out by the directive. (An assessment for P reduction has not yet been made.) This study also carried out an assessment of the cost effectiveness of implementing such a reduction and established costs at €13 million pa which equates to between 0.5 and 0.6% of the total production income for the catchment area, and thus very achievable. This would involve the re-establishment of wetlands, use of catchcrops and a reduction in fertiliser use, All beneficial to the local wildlife. The author notes, however, that current and future data will certainly be influenced by nutrient loss in the Baltic catchment as a whole (Petersen *et al.*, 2009).

5.6   Part 4: Aquaculture/fish farming

Worldwide, fishing is in trouble. Less fish and smaller fish are being harvested from the ocean. There is no real way to quantify the decline in fish stocks, but estimates are not generally encouraging. To try to alleviate this problem many countries have introduced fish quotas and in some areas 'no-take' zones have been imposed. Such measures have been met with mixed feelings, and many fishermen whose families have been in the trade for generations have given up, sold their quota and moved on to pastures new. However, these measures are valid and, although they may not strictly meet many people's idea of 'farming', they do constitute a degree of livestock management which is relevant in the context of conservation within farming practices. This subject is therefore discussed later in this chapter.

History is replete with tales of ghost towns created by 'boom and bust' fishing practices, for example famously, the anchovy fishery along the Peruvian coast. The Peruvian Anchoveta Fishery took off exponentially from the mid-1950s, whole communities developed along the coastal area all dependent on this one commodity. The industry built up to a peak in 1970 with a catch of over 12 million tonnes that year. However, by 1972, the fishery encountered an almost total collapse and the local fishing villages became deserted. El Nino has been partly blamed for this collapse; however, the fishermen consistently took up to 2 million tonnes in excess of official recommendations (Webber and Thurman, 1991; Cousteau, 1975).

With rising demand for food fish and fishmeal, and an unfathomable decline in wild fish stocks we need to look elsewhere for answers. Aquaculture is one way forward and is currently on a rapid incline with respect to supplying this growing demand. It now provides approximately 40% of the world's seafood production (Langan, 2008, personal communication). It is by no means new, the Chinese have been using aquaculture for centuries and some of their practices are now being revisited by scientists and stakeholders. As with all farming and indeed with all industries that are 'on the up' there are problems to be faced.

It seems that having anything positive to say about fish farming stirs up a frenzy of stern rebukes from academics and environmentalists alike. So finding ways to make improvements in farming techniques that will reduce these impacts and improve production is high on the agenda. This approach seems to be developing quite well with other farming disciplines so why not here? Our research has indicated that there are numerous producers with a very proactive environmental approach to their operations. From some of the larger companies around the world, to some much smaller operations, whether by legislative pressure or, more often than not, by choice, some often novel approaches can be observed with the aim of reducing environmental impact.

In this part of the chapter we will address this hot topic head-on as follows:

- A potted history of aquaculture
- Other uses for aquaculture products
- Case Study 5.2 Dragon Feeds Worm Farm, South Wales
- Aquaculture and the Environment
- Case Study 5.3 Loch Duart Salmon Farm, Northern Scotland
- Case Study 5.4 Fisheries in New Hampshire, US
- Some other examples
- The Future of Aquaculture.

5.6.1   A potted history of aquaculture

Aquaculture is the farming of aquatic organisms, both fresh water and marine, usually for commercial purposes, as with other forms of farming, or sometimes for scientific and conservation purposes such as captive breeding programmes. Usually aquaculture is used for finfish, crustacea and molluscs though sometimes also amphibians and, of course, aquatic plants.

Harvesting of wild marine and freshwater organisms, or fishing, has a history that certainly precedes any formally kept records, but surprisingly so does aquaculture. It is believed that the Chinese have been raising food fish in their paddy fields for well in excess of a thousand years (Bocek – Auburn University, Alabama, US – accessed May 2009), using a system now known as polyculture. Polyculture is the raising of two or more compatible organisms together within an aquaculture system; the different species in the system complement each other and do not compete. They may live in different levels of the water column, and have different feeding niches, for example grazers, invertebrate feeders, detritovores, and so on. In this case, the cultivated rice plants would attract insect and mollusc grazers, carp being reared in these paddy fields would feed on these, reducing predation on the rice plants and at the same time fertilising the system. Without the fish, the farmer would need to add nutrients and manage the grazers, likewise a farmer trying to raise the fish in such an enclosed system without the rice, would have wastes to dispose of, and may have a problem with levels of Biological Oxygen Demand (BOD), this is the O^2 used by bacterial decomposers. The farmer works with a system that partially takes care of itself, and gets both a rice crop and a fish-catch into the bargain. As in this example, the raising of both plant and animal within a polyculture system is also sometimes known as aquaponics.

How much the ancient Chinese understood of the biological cycling of nutrients and photosynthetic/respiratory gases is unclear, however what is clear is that they were able to recognise that the closer they mimicked a natural system, the more fruitful their harvest. China still produces around two thirds of the world's aquaculture products.

Ancient Hawaiians also carried out a form of aquaculture; they built low rock walls out from the shore thus creating fishponds. The fishponds at Molokai are a popular tourist destination. The ancient Hawaiians also developed freshwater fishponds in upland regions. These operated as polyculture systems. Taro plants have an edible root and leaves when cooked. These were grown in mounds in the ponds. The fish, mainly mullet and perch would feed on insects predating the taro plants. They also fed on some of the leaves, but this actually allowed the plant to develop better by removing apical dominance in the 'pruned' plants (Rosauer, 1987). Also found on a rocky shore on Heisker, the Monach Isles National Nature Reserve located off North Uist, Scotland and some 40 miles from St Kilda, Scotland, where the residents built a 'cairidh' or-'ghari' Gaelic for fish traps. The tide would bring in shoals of fish that would become trapped behind the wall as the tide fell and the women would scoop up the fish in their aprons (Sutcliffe, 2010, personal communication).

Archaeological evidence and cursory mentions in ancient writings indicate that many other ancient cultures including the Japanese, the Egyptians and the Romans carried out aquaculture of fish, molluscs and seaweeds, though detailed records are limited. However, it wasn't until modern times that aquaculture practices became better documented.

5.6.2   Other uses for aquaculture products

Farming for food fish and shellfish is not all that aquaculture has to offer. Other industries have arisen with other products required or desired by consumers such as cultured pearls for use in jewellery. Also cultured microalgae and macroalgae (seaweeds) are used in many industries from bioplastics to the cosmetic industry. These in themselves are outside the remit of this text, however, as an adjunct to the fish farming industry, another production system was brought to our attention during our research. Tucked down on the south coast of Wales an entrepreneurial industry has arisen farming ragworms as a replacement for fishmeal. With an ever increasing demand for farmed fish and therefore fish feed, and an uncertain future for wild catch used for this fish feed, this industry may prove to be a way forward. Case Study 5.2 below looks at this small but growing company, Dragon Feeds Ltd.

Case Study 5.2: Dragon Feeds Ltd

Dragon Feeds Ltd has been established since the 1970s. In fact during our visit to their operation in South Wales, Managing Director Tony Smith proudly announced to us that he has been 'in worms' for 30 years! With the demand for farmed finfish and shellfish increasing on an annual basis, new methods of production are constantly on the agenda, and therefore so is the demand for feed.

The company has a novel approach to the production of fish food by using polychaete worms (*Neiris virens*) to replace fishmeal in their feedstuffs. This is a very natural and attractive food for fish, indeed people have been using these worms to bait fish, probably for as long as people have been fishing.

Fishmeal

Fishmeal is a high quality product widely used in aquaculture and in livestock feeds (especially pig-feed and chicken-feed), as well as horticulture to a limited extent. Feeding fishmeal to ruminants was banned in 2001 in the wake of the Bovine Spongiform Encephalopathy (BSE) scare, however it was reinstated in 2008 (www.gafta.com/fin – accessed June 2009), and there is some evidence that bovine fertility is increased by its use (Staples *et al.*, 2009). It has a wide range of very specific nutritional components, mainly proteins with a very rich amino acid compliment, oils, fibre, minerals and vitamins. It is simply made using both whole fish and trimmings from food fish. These are cooked, pressed, dried and milled into a powder form. This product is highly digestible compared to plant-based proteins from, for example soyabeans that are also commonly used in feeds (Olli and Krogdahl, 2008).

On an international basis, most fishmeal consists of small, bony pelagic fish, taken specifically for this purpose. In the EU only around a third falls into this category, the remainder consisting of food fish trimmings – however in the UK, France, Spain and Germany the proportion of trimmings used is much higher (www.gafta.com/fin – accessed June 2009). In South America fishmeal is made mainly from anchovy and horse mackerel, North America mainly from pollock and menhaden, Scandinavia mainly herring, capelin, sand eel, sprat and blue whiting, South Africa uses mainly the pilchard. The rest of the world uses a wider variety of fish for their fishmeal. This is therefore a huge component of the world fishing industry.

These fish must be processed rapidly as the rate of decay is high, so many companies use factory ships for this purpose. Where processing occurs on land, very strict environmental monitoring is required as effluent from processing can raise the BOD to unacceptable levels (www.gafta.com/fin – accessed June 2009).

There is already concern amongst the industry and the public at large about the health risks associated with pollutants accumulating in fish and shellfish tissues. As a result strict legislation exists in many countries to check quality control systems used by the industry with respect to fish entering the human food chain. However not all nations show the same concern with respect to fish used in fishmeal, though processing may well reduce any potential threat, very little data is available to indicate whether this has been fully analysed.

Coupling this with the simple fact that worldwide fish stocks are already in decline and that the demand for fish for fishmeal as well as for direct consumption is increasing to an unknown degree, perhaps it is time to start thinking about alternatives to supplement this important feed component.

Polychaetes as a Basis for a Sustainable Alternative to Fishmeal

The specific biochemistry of fishmeal makes the production of sustainable replacements a complex issue. Dragon Feeds Ltd has a highly qualified and experienced team carrying out ongoing analyses of these components and is constantly trialling new products to meet an increasing demand for a replacement to fishmeal.

The main component is the polychaete worm and these are farmed on site in large open tanks of seawater. One polychaete female produces approximately 7×10^6 eggs, and the farmed offspring have a very high survival rate, around 80%. The worms are the basis of the feeds and are high in protein, fatty acids and vitamins. These are combined with other ingredients such as soya proteins, and algae, depending on the nutrient requirements of the customer. The end products are in pellet form. These pellets are used for farmed finfish, for shrimp production and for the ornamental market, and different feeds are developed that are specifically formulated for each of these markets or can be tailor made where requested.

Tony Smith described their production system as a 'gentle' process, designed to reduce potential denaturation of the polychaetes proteinacious components – essential amino acids being the prime nutrient requirement derived from the worms. Production and processing of the polychaetes, plus information on all of the components in the feed, which have been analysed down to trace element level, is freely available on Dragon Feeds Ltd website (www.dragonfeeds.com).

The resulting product is sustainable, traceable, contains all required nutrients and, the inclusion of the worms not only contributes significantly to the nutriment balance, but also makes the product highly palatable to the fish, plus there is the added benefit that end consumers do not have to be concerned with the possibility of contaminants that may be associated with fishmeal.

Fish Farming Using Dragon Feeds Products

Having produced small numbers of fish over the years for product research and development purposes, the farm has now started to expand its own fish production using purely its own feed products. They now have large seawater ponds where they produce sea/brown trout, rainbow trout (*Oncorhynchus mykiss*), bass (*Dicentrarchus labrax*) and bream. These large raised ponds hold in the region of 2400 fish each. The palatability of the feed was clearly evident at feeding time, we were there when the trout were being fed and these fish were vigorous and healthy and ready for market.

Sustainability and Clean Water

Seawater for the beds comes from the nearby bay; this is subject to the tide, and ranges from 20 l per second at high tide down to a trickle at low tide.

As well as providing the main component of this feed alternative, the polychaete worms also perform a service as detritovores within the tanks helping to maintain water quality. Extraction water is regularly tested by the EA, and output is generally in the region of 27× clearer than input water.

5.6.3   Aquaculture and the environment

Aquaculture is globally notorious for the damaging effects it has on the environment. Fish farming especially involves production of highly energy demanding species in an enclosed system, often causing major problems with respect to waste disposal.

The farming of salmonids such as salmon and sea trout represents some of the largest aquaculture operations in the UK and in North America and Canada. Owing to the size of these operations, their environmental impact has been studied extensively. A relationship between organic enrichment of the benthos in the immediate area of a number of Scottish salmon farms, and the currents in those areas has been observed as has a correlation between increasing levels of hydrogen sulfide and fish mortality in these areas, though this would require a high energy input, and even then the hydrogen sulfide would have to have accumulated over an extended period of time (Black, Kiemer and Ezzi, 2007).

Another environmental impact has been raised with respect to salmonid genetics. A high proportion, possibly as much as 94%, of all Atlantic salmon (*Salmo salar*) are captive bred in aquaculture systems. It has been proposed that captive bred Atlantic salmon are evolving to be genetically distinct from wild Atlantic salmon (Gross, 1998). If this is the case, then the long term implications to the wild fish gene pool cannot be predicted where escapees have found their way into the wild.

Furthermore there is no question that one particular species of sea lice (*Lepeophtheirus salmonis*) is commonly transmitted between captive and wild salmon stocks. One report from British Columbia indicates that where salmon farming is especially intense, up to 80% of wild salmon may become mortally infected (Krkosek, 2005). There has been some research to examine whether wrasse can be used as cleaner fish to remove these parasites, but little progress appears to have been made at this time.

The importance of salmon fisheries cannot, however, be overstated and the pressure is on for fish farmers in the industry to reduce their impact on the environment, and legislation is in place to enforce this to a degree. In the North West of Scotland we found a salmon farmer with a whole new approach and with a sound philosophy applied to their production system. Loch Duart Salmon Farm is the subject of Case Study 5.3 below.

Case Study 5.3: Loch Duart Ltd Scotland

It is a simple fact that wild Atlantic Salmon stocks are dwindling, however the demand for salmon is increasing on an annual basis, the necessity for salmon farming is therefore also on the increase. In 1999 salmon farming tonnage was, for the first time, higher than caught wild salmon tonnage, and the proportion of farmed versus wild salmon has increased each year since. When we think of salmon farming an image of high intensity, environmentally damaging production springs to mind. Legislation exists to try to ameliorate this impact as discussed above, but we found a salmon farmer whose whole philosophy on salmon production could change the way that we look at this industry.

Loch Duart Salmon based in Sutherland, North West Scotland, leased from the Crown of State and on South Uist, in the Hebrides, uses a production system which not only dramatically reduces environmental impacts, but also produces fish which are not stressed, but are very healthy, 'happy' fish and which are also healthier to eat, as toxin levels in the tissues are minimal. Potential health problems arising from eating fish containing polychlorinated biphenyls (PCBs), dioxins and mercury accumulated in their tissues has been well documented and recommendations for maximum salmon consumption varies from total abstinence to three to four times per week, however Managing Director Nick Joy states that he is confident that you could eat Loch Duart salmon every day.

Production Manager Mark Woods very kindly took time out to show us around the company's operation near Scourie, Sutherland (shown on Map 5.1). The difference between this farm and other salmon farms is immediately apparent. Their approach is to produce high quality farmed salmon that is alike to wild salmon by mimicking natural life cycles, feeding regimes and shoaling densities as closely as possible. At the same time they are constantly monitoring the fish and the surrounding environment.

Salmon Life Cycle within the Farm

Loch Duart runs its own brood stock programme. Brood stock are kept at sea until the end of the harvesting period, around November to December, when they are moved to fresh water where they are stripped of eggs and milt mostly for the company's own use, but some go to third party suppliers to bring on. Atlantic Salmon are anadromous, that is they are born in fresh water but live most of their life cycle in the sea before returning to fresh water to spawn, so in the wild they would be spawning in the fresh water sites, usually where they hatched, at this time of year, the timing in the farm therefore closely mimics the natural life cycle of the fish. The fertilised eggs are laid down in the hatchery. The company has a brand new hatchery, opened in 2008 and designed in conjunction with the Royal Society for the Prevention of Cruelty to Animals (RSPCA) Freedom Foods initiative.

Eggs are sorted when they have developed into 'eyed eggs' that is when the black eyes spots can be clearly seen. Those without these spots are unviable and are discarded.

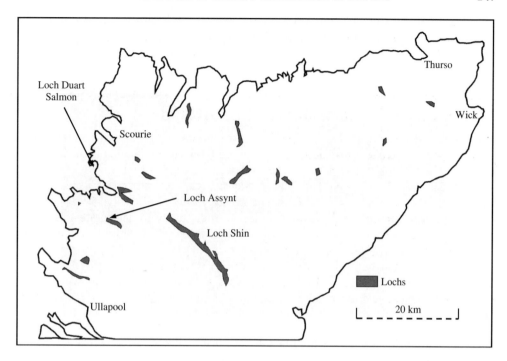

Map 5.1   Site of Louch Duart Salmon farm, Scotland

Hatching occurs between January and April, at which stage the hatchlings, called alevins, are checked for disease, any diseased specimens are discarded and the viable, healthy stock is placed into $6\,m^3$ tanks. These tanks are heated to $2°$ above ambient. Alevins live off their yolk sacs in the first instance but as they use up this store of food external feed is introduced, at this stage the young fish are known as 'fry'. These fry are grown on in these tanks to approximately $1\,g$ in weight, at which time they are able to adjust and hold their position in the water and they become known as 'parr'. They are graded according to size and are vaccinated against disease by hand at this stage. As the parr develop and grow in size to approximately $2\,g$ in weight they are transferred to the new larger tanks, (Figure 5.6).

These new tanks are $5\,m$ in diameter and $2\,m$ deep, they have two gravity fed inlets, one with ambient water, the other with saturated water when more oxygen is required. Oxygenation is mainly by hand, but with fine control from the computer system, which constantly monitors the O_2 levels in the water. Further oxygenation is supplied by airstones placed in each of the tanks. All the tanks have two independent pipes, one for the transfer of the fish between tanks – this reduces handling stress, the other is an outlet pipe for waste water. These outlet pipes have a $60\,\mu g$ filter to remove solids, the water is not circulated at all, so it is a single pass system.

In March to April of the second year the fish smoltify. Physiological changes occur which prepare the fish for life at sea. This process is easily identified as their appearance also changes at this stage, the colouration becomes silver with a white belly and at this stage they are transferred to open water pens.

Figure 5.6   Salmon rearing tanks at Loch Duart. (Photo: Sarah Burchett)

Sea Pens

The Scourie site has nine sets of 14 pens, each $15\,m^2 \times 10\,m$ deep and positioned in 20–30 m of water. The positioning of the pens has been chosen by the company marine scientists, who have made a study of the local hydrology. Water movements have been carefully analysed to ensure that there is no build up of potentially harmful wastes. This helps to maintain both the health of the fish and of the seabed environment. Loch Duart's unique fallowing system ensures that only six of the nine sets of pens are in use at any one time, the other three lay fallow for a full year in turn, most salmon farms only have a six week fallow period. The fallow year ensures the health of the seabed, and reduces the possibility of disease outbreak by breaking the disease cycle. Of the operational sets of pens 2 of the 14 will always be empty. These are raised above the water to dry out, this means that no antifouling chemicals are required to clear the nets of marine growth, further helping to maintaining the health of the seabed and the pen environment (Figure 5.7).

Fish are cycled between pens using pipes that they swim through so as to reduce handling stress. Each pen contains approximately 7–8000 fish, roughly half the standard capacity. The fish are able to move around, grow naturally and develop properly. They are sleek, healthy and have powerful fins, making it difficult to distinguish them from their wild counterparts.

Harvesting

The smolts take one to two years to grow into adults; they are graded according to size, which is naturally quite variable. Some salmon mature after the first year, these are known

Figure 5.7   Nets are raised out of the water regularly to dry out. This kills off primary colonisers and removes the need for anti-fouling chemicals. (Photo: Sarah Burchett)

as 'grilse' and these will be harvested at this stage. Most salmon mature after two years. Harvesting is carried out throughout the year when the fish reach the desired weight of 4–5 kg, fish colour and flavour has natural variation due to the natural way in which they are grown. Older fish are generally sent to the smokeries. Pen stocking densities are maintained as the fish grow by harvesting and size grading.

Feed

The salmon are fed on the 'Loch Duart Diet' that mainly comprises herring trimmings, Loch Duart is very particular about the source of this meal. Products from fisheries with a high bi-catch record are not used, they use a scoring system to monitor the environmental impact of all the materials they buy in. Feed distribution is undertaken to ensure that there is no competition between the fish in the pen and that the fish on the perimeters receive their quota, staff are trained in 'appetite monitoring' and periodic starves, of no more than 72 hours at a time, are used, again to mimic natural feeding regimes.

Disease

The main disease issue with salmon is lice. Just a few lice are enough to kill a salmon and they can spread rapidly. Salmon farms in Chile have suffered major lice outbreaks due mainly to high fish densities within the farms and to the close proximity to other farms. Loch Duart tackles this in a number of ways. Constant monitoring throughout the life cycle, cycling of treatments to reduce the possibility of resistance, and of course the long fallow period which breaks the disease cycle. *Furunculosis* is a bacterial disease still common in wild salmon but controlled in farmed salmon by inoculation. Salmon can also be infected by viruses and fungal diseases. At Loch Duart, the fish are checked visually on a daily basis to look for any signs of poor health, any dead fish are removed and any sick or injured fish are removed and humanely killed. Water quality in the pens is also monitored daily by taking temperature, turbidity and oxygen (O_2) measurements. O_2 is not allowed to fall below 6 ppm.

Environment

Stringent biosecurity measures are in place. Staff are required to use Personal Protective Equipment (PPE) at all times when working on production sites, and this equipment is disinfected at every movement between sites or areas within those sites, anti-bacterial gel is used at the freshwater sites. We, as visitors, were required to do the same.

Looking after the health of the fish does not just make good economic sense; it is also an essential component of maintaining the health of the local environment. Where there is disease, or where there is overstocking, thus increasing waste products, the surrounding water quality and the seabed suffer the affects too.

Loch Duart has put in place many procedures to address any potential damaging affects of their production on the environment. A number of these have been discussed above:

- No anti-fouling chemical is used to clean the pen nets, instead each net is raised clear of the water on a regular basis to kill off any marine growth accumulating on it.

- Low stocking densities reduce the waste products.

- Regular starve periods, mimicking natural feeding regimes, also have the benefit of reducing the potential for unconsumed feed rotting in the water.

- A full year of fallowing for the pens in each of the three sea lochs alternately to allow for seabed recovery and to help break the disease cycle.

- At the freshwater sites, loch water goes through the pens in a single pass system, and has a fine particle filter to remove solids, no overfeeding of fish prevents nutrients passing into the loch at the output and causing nitrification of the water.

- Ongoing monitoring of fish health, water quality and abiotic factors, visually, biochemically and by computer systems, is carried out.

Further to these precautions, the company is pioneering a unique polyculture system. Still new, and subject to further research, this system has huge potential for the future of fish farming as a whole. Loch Duart has introduced urchins and seaweeds to pens on their sites. The system works as follows:

- The Black Urchin (*Paracentrotus lividus*), is a native species that can live on any leftover feed and on algae attached to the sides of pens. These therefore help to keep the pens and the immediate environment clear of potentially harmful products and helps maintain water clarity. The urchin can also be marketed, which, although not popular in the UK, is considered a delicacy in other parts of the world especially France. The urchins must, however, be removed at the end of each cycle otherwise they could potentially become a biohazard in themselves as they would move elsewhere to feed. They must also be graded by size as they have a tendency for cannibalism.

- The seaweeds, Dulse (*Palmeria palmate*) and Sugar Kelp (*Laminaria saccharina*), both native species, take up the nutrients from the waste products of the fish, and they also sequester carbon from the surrounding environment. These too can be harvested as they represent a useful source of minerals, and dulse is often eaten as a delicacy.

The potential of this system is for nutrient 'balancing', that is, as many nitrates and phosphates are being removed from the seas as are being created by the fish farming process.

Biological monitoring of the loch seabeds and of the freshwater lochs indicate very little environmental impact. In the sealochs, as well as the indigenous sessile life, many species have been recorded including dolphins, porpoises and occasionally minke whales. Seals are often in and around the lochs looking for an easy salmon meal. Every effort is made to protect the fish from the seals using non-harmful methods. The pens have outer-perimeter nets (predator nets) that constitute 'curtains' down to the seabed 20–30 m below. Acoustic devices are also used to deter the seals.

Community

Loch Duart's philosophy also incorporates maintaining a relationship with the local community. They employ around 60 staff many of whom are locals. They liaise with nearby schools and supply work experience programmes and school trips. Staff are encouraged and supported at all stages to expand their skills in such pursuits as seamanship, safety awareness, data collection and husbandry.

Markets

With an annual turnover of approximately 3000–4000 tonnes, the company has approximately 2.5% of the market share of Scottish produced salmon. Around 30% is marketed in Scotland and the rest of the UK; around 20% goes to Europe and a further 20% to the US market. The remainder is sent to various other parts of the world including Asia. A number of top chefs have specifically recommend Loch Duart Salmon, and prepare it in their restaurants. Despite the low stocking densities, this indicates that the quality of fish produced in this environmentally sound manner does not appear to have a negative effect on company profits, on the contrary it indicates that a high quality product attracts a premium price and consistent demand. Something to be considered perhaps by companies struggling to comply with strict environmental legislation.

Certifying Bodies

The company is certified with the following:

- Fisheries Research Services (FRS)
- Food Certification Scotland (FCS) Label Rouge
- FCS Code of Good Practice (CoGP)

- RSPCA Freedom Food Welfare Standard
- Scottish Environmental Protection Agency (SEPA)
- Investors in People (IIP)
- Has committed to selling only Marine Stewardship Council certified fish by 2011.

Loch Duart is also certified with the ISO 14001 for its standards in Environmental Management Systems and all staff are trained in its implementation. The company's Environmental Policy Statement can be read on their website, www.lochduart.com as can details of the above.

With wild salmon under threat, the production of farmed salmon can only increase to satisfy demand and thus to help protect wild stocks by limiting commercial salmon fishing. If environmentally sound salmon farming can be done in a commercially viable manner, and the above case study indicates that this is possible, then the future for wild stocks of Atlantic Salmon may not look as bleak as once anticipated.

Another approach to reducing the impacts of fish farming on the environment, and to ensure the improved welfare and product quality in the farmed fish, is to take the operation way offshore. We found a number of operations that have taken this approach. Although these are generally in the early stages of development and undergoing some teething problems, highly motivated and skilled teams of fish farmers and scientists are smoothing their way into the future of aquaculture. Case Study 5.4 below is an example of this work from New Hampshire, US.

Case Study 5.4: Offshore Fisheries in New Hampshire, US

In the US, the University of New Hampshire's Marine Aquaculture Center has been developing techniques for raising native coldwater finfish and shellfish out at sea where the environment is much as it would be were they living in the wild. Most seafood production currently in aquaculture is carried out in protected bays and inshore waters. Such areas engender inherent problems. For example, these are often busy areas used for recreation and conventional fishing and are therefore already subject to environmental pressures. Also, tidal movements are often not sufficient to effectively remove waste products from large volumes of farmed fish. Moving such production out to the open ocean would seem to be a natural progression.

Dr Richard Langan, director of the Atlantic Marine Aquaculture Center, took us 6 miles (10 km) out into the Gulf of Maine to see their pioneering trial operation. On a 30 acres (12 ha) offshore site Dr Langan and his team have set up a high-tech system of cod (*Gadus morhua*) production in four large sea cages, or pens, which are 3000 m^3. These cages at the time of our visit were made from woven netting and, as there can be problems with potential damage to these nets, the team are currently looking at replacing these with cages made from more durable materials. Mussels on longlines are also being trialled on the site. Figure 5.8 shows a schematic of the system.

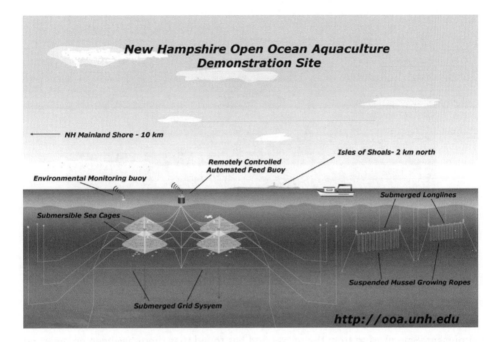

Figure 5.8 Schematic of the New Hampshire open ocean aquaculture demonstration site. (New Hampshire Open Ocean Demonstration Site, with kind permission of the University of New Hampshire)

Here fish develop in a near-normal environment, with each pen holding 120 000–145 000 individuals. Natural water flow from tidal movements and currents maintain the habitats' health, cod are shoaling fish so would normally be found in large collectives such as those recreated by this system. The sea cages are monitored by a station located at the site (Figure 5.9).

Feeding is remotely controlled from a specialised feed buoy, with a delivery of 1% of the pen weight once per day. Monitors in the on-site station also observe behavioural patterns. Any fish showing signs of potential disease, plus any dead fish, are removed quickly by divers. This method of recovery will not be practical in the long term, especially when the system becomes scaled up for commercial use, not just because of the practical logistics of using divers but also for health and safety reasons. We were fortunate to have a calm day for our visit, but weather patterns are unpredictable, and conditions are not always suitable for diving, so ways of mechanising these routine operations will need to be developed. The monitors are both solar powered and have diesel generators as a back up. The station has a cellular link to the internet where the pens may be observed remotely. We joined the team on the station to see the operation in progress. We found it high-tech, but relatively simple to operate and with efficient back ups, all essential considerations if this system is to be adopted on a wider scale.

As with all research there are some issues to be ironed out (this is by definition what the trials are for). We put a number of pertinent questions to Dr Langan.

We wanted to know just how environmentally friendly this system is, what has been the impact on the immediate ecosystem? He told us that the centre has been carrying out

Figure 5.9   Cod hatchery monitoring station. (Photo: Sarah Burchett)

environmental monitoring from the outset and has found that there has been no measurable impact on the surrounding ecosystem.

The environmental impact is of course one of the key parameters that is considered by the state authorities when an application is made for issuing aquaculture permits. Dr Langan told us that obtaining such permits for expanded operations will be difficult until a regulatory framework is in place for siting farms in federal waters. Large operations for either finfish or shellfish in state waters (<3 nautical miles) would likely run into conflicts with fishing and other activities. With the regulatory uncertainty, securing investment will also be a problem.

The fish that are currently being farmed, albeit on a small scale for research purposes, are marketed. However rolling this system out into the wider commercial world is likely to engender its own difficulties. We put it to Dr Langan that setting up such an operation so far offshore, and putting in the infrastructure to maintain it, is likely to involve considerable up-front investment; that is quite a risk for a commercial company. Dr Langan explained that while more costly than nearshore farms, at the proper scale and with the right species, he believes that offshore operations should be profitable. There is some interest from industry, but said Dr Langan, 'until the salmon and bass and bream farming companies make a commitment to move offshore, growth of this sector will be measured'.

It becomes clear that, though there are some difficulties in moving forward with this project, he and his team are highly committed to getting this exciting project off the ground in the near future. The success of this type of research, and the future of its sustainable commercial application, is important for the environment, for the health and general welfare of the fish and for the quality of the product.

The use of offshore sea cages could become a common theme for the future and is further considered below in the sub-section on the future of aquaculture.

5.6.4   Some other examples of conservation within fish farming

Polyculture, whereby the raising of two or more compatible organisms together within an aquaculture system, was discussed above with respect to the historical origins of fish farming, but polyculture is used today by many nations. In modern times it is being recognised as key to sustainable aquaculture as such methods, at least partially, mimic natural multi-trophic systems.

- A visit to Malaysia in spring 2009, revealed this type of sustainable polyculture in action. Rice paddies near Kuala Selangor to the west of Kuala Lumpur, have been planted alongside a river. The river has been diverted to encircle the paddy fields (Figure 5.10) to maintain flushing of the crop with fresh river water. This water is then carried out back into the main stream of the river. Here holds have been placed to raise various species of river fish, which consume run-off products from the paddies (Figure 5.11). Without these fish, eutrophication may, occur causing long-term damage to the river ecosystem.

- It is a simple thing to define farming as the raising of livestock and the cultivation of crops for food, fuel or fibres. It could, however, be argued that any strategy used to manage plants and livestock, whether domestic or wild, for our own uses should also be described as farming. Below are a number of examples where

Figure 5.10   Malaysian rice paddy. (Photo: Stephen Burchett)

Figure 5.11   Rice and fish farming by polyculture in Malaysia. (Photo: Stephen Burchett)

management of fish stocks have proved beneficial to the future of fisheries and to conservation.

- Goat Island Marine Reserve in Leigh, New Zealand lies some 55 miles north of Auckland, and was one of the first of its kind. It was the result of a 12-year battle by an Englishman Bill Ballantine who moved to New Zealand to run Auckland University's Marine Laboratory. The reserve opened in 1977, despite the protests of many local people, especially fishermen. The recovery of the underwater life in the $5\,km^2$ site proved to be so successful that within a few years, fishermen had bumper harvests in nearby waters from the spill-over of eggs and larvae from both mature fish and from crustacea which originated in the reserve, particularly snapper (*Lutjanus spp.*), crayfish (*Jasus Ialandei*) and spiny rock lobster (*Jasus edwardsii*). The area was managed specifically for conservation purposes and proved to be a resounding success for both its primary purpose and for local fishermen and is used extensively for education and research, thus benefiting the community as a whole (New Zealand Geographic, 2004).

- The Negev Desert in Israel has a rainfall of <100 mm per annum and seems an unlikely place to find aquaculture, however, since the early 1980s fish have been farmed here. Oil drills have been used to tap geothermal water from deep aquifers under the desert sand. This water is slightly brackish (approximately 10% the salinity of seawater or just over 3‰ salt), and has a temperature of around 30 °C. It is high in minerals and has low pollution and generally low O_2 levels. Tilapia are particularly well adapted to cope with these conditions and are therefore one of the main types of fish farmed in this way; the conditions are also good for primary production which forms the base of the food chain. Though the geothermal water is of poor quality for agriculture when it is first extracted, it has proved to be of use after cycling through the aquaculture system. For example, a small fish farm capacity of approximately 2000 m^2 refreshes its water at a rate of approximately 10% per day – or 200 m^2. This water will now be higher in nutrients and lower in minerals after passing through the aquaculture system and can be used to irrigate around 4 ha of crops, such as olives and tomatoes. There have been a number of logistical problems to be ironed out over the past three decades, but the current systems are efficient and productive, and Israel has passed on this technology to other desert nations (Kolkovski *et al.*, 2003).

- Situated on the Eastern Coast of North America flanking the states of Maryland and Virginia is Chesapeake Bay, the largest estuary in the United States. This area is especially famous for its oysters. Eastern Oysters, also called Atlantic Oysters (*Crassostrea virginica*) are a key species in the Chesapeake Bay ecosystem, accumulating in vast numbers and forming reef-like sub-strata, a habitat on which many other organisms, especially invertebrates like Porifera (sponges) and Tunicates (sea squirts) depend. Their juveniles are also an important food source for other animals such as blue crabs (*Callinectes sapidus*) and some fish, and they provide a highly efficient natural filtration service that helps to maintain the health and clarity of the water, each filtering up to a staggering 50 US gallons (approximately 190 l) per day. These oysters have been harvested for around 400 years and have been an economically important industry for the area for about the last 150 years (www.chesapeakebay.net – accessed January 2010). Unfortunately they have been over-fished and numbers diminished to the point where the ecosystem is seriously under threat. Now aquaculture companies such as Marinetics Inc. – The Choptank Oyster Company in Maryland are farming these native oysters. This reduces pressure on wild stocks and may indeed help to replenish wild stock numbers as surplus spawn is utilised for restoration programmes. Unfortunately some researchers indicate that such restoration may be too little too late as a high percentage of the areas selected for restocking are on seabed that is too silted for larval settlement (Smith *et al.*, 2005). But it remains to be seen whether this means that the process is either untenable or just more long-term than previously considered. Either way, oysters produced by aquaculture must certainly relieve the pressure on wild stocks, along with the implementation of 'no-take' zones by the Maryland State Government; and will continue to help supply biological services by filtration in the bay.

- Another example where natural marine resources are managed to increase harvests come in the form of 'fish boxes'. In Norway plaice boxes have been used for the last 20 years. These are areas along the coastline of the Wadden Sea, uniformly shaped (hence 'boxes') for easy identification by the fishermen. These boxes are favoured by juvenile plaice. In the boxes beam trawling is restricted, or in some cases completely banned, allowing the fish to develop and grow, thus increasing stock (The Seafish Organisation, 2007). This may not be relevant to many species, as not all fish have grounds specifically favoured by their juvenile populations, but it is another good example of long-term conservation management strategy coupled with more sustainable fisheries.

5.6.5   The future of aquaculture

The use of sea cages was discussed above in Case Study 5.3. In Hawaii, a similar project is underway. The production system is carried out in huge aluminium and Kevlar™ spheres called Oceansphere™. These spheres measure 162 ft across (approximately 50 m) and are untethered in the sea. The sphere system can yield as much as 2000 tonnes of biomass, in this case mainly yellowfin (*Thunnus albacares*) and bigeye tuna (*Thunnus obesus*), in less than half of a square mile. The Oceansphere™ is powered by a patent pending hybrid ocean thermal energy system that does not depend on fossil fuels, thus reducing environmental impact (Spencer and Troy, 2009, personal communication). Bill Spencer and Paul Troy who are heading up this project inform us that the practicalities of the system are clearly thought out. The spheres are modular and are therefore easy to repair, the start-up costs for investors is relatively low considering the returns, and, because of the size of the spheres, the stocking rate to volume ratio is not very high, reducing both the environmental impact and maintaining fish health. As the spheres are placed in deep water the sea temperature and the currents remain relatively constant and the dilution of wastes is increased. As they are untethered, benthic accumulation is significantly reduced.

As discussed above, there are many conservation issues to be considered with respect to aquaculture practices and the environment. One of the objectives of this book is to look at some of the measures that producers are trying to put in place to address these issues. It must be recognised that aquaculture is important to our future as a species with respect to food security and socio-economics. As such, although we recognise and acknowledge the concerns of environmentalists, this book aims to look at the bigger picture and communicate our findings where we can see that hard working and committed stakeholders in the industry are making concerted efforts to alleviate these concerns. We consider communication to be the key. Across the world, good aquacultural research and development is under way, we have touched on only a fragment of this work, but there is so much more to be reported and new methods to be gleaned. Journal articles have their place, but such information needs to be disseminated to end users. We found one such project in Africa. The 'Sustainable Aquaculture Research Networks in Sub-Saharan Africa' (Sarnissa). Funded by the EC, and supported by African, European and Asian universities and non government organisations (NGO's), the project aims to illustrate sustainable aquacultural methods that also address growing food security issues in a number of African countries. The project is developing an 'Aquaculture Compendium' for use by stakeholders as well as giving open access to relevant journals, and offers a significant platform for communication and research into the future, accessed through their website on (www.sarnissa.org – accessed July 2009).

Humans have always naturally gravitated towards water. It is an essential component in our lives, everything we do requires it, almost everyone on the planet lives a stone's throw from a watercourse, a pond, a lake or the ocean, and its importance to us, and to all other living things, cannot be overstated. We must attend to its health at all costs for all our sakes.

References

African-Eurasian Waterbird Agreement (AEWA) website – www.unep-aewa.org.

Asher, J., Warren, M., Fox, R. et al. (2001) *The Millennium Atlas of Butterflies in Britain and Ireland*, Oxford University Press, Oxford.

Black, K.D., Kiemer, M.C.B. and Ezzi, I.A. (2007) The relationships between hydrodynamics, the concentration of hydrogen sulfide produced by polluted sediments and fish health at several marine cage farms in Scotland and Ireland. *Journal of Applied Ichthyology*, **12** (1), 15–20.

Bocek, A (ed.) *Water Harvesting and Aquaculture for Rural Development*, International Center for Aquaculture and Aquatic Environments, Swingle Hall, Auburn University, Alabama 36849-5419 USA, http://www.ag.auburn.edu/fish/international/polycul.htm (accessed May 2009).

Cousteau, J.I. (1975) *Riches of the Sea*, Angus & Robertson.

Gross, M.R. (1998) One species with two biologies: Atlantic salmon (*Salmo salar*) in the wild and in aquaculture. *Canadian Journal of Fish Aquatic Science*, **55** (S1), 131–144.

Jackson, D.L. and Jackson, L.L. (2002) *The Farm as a Natural Habitat*, Island Press, Washington, DC.

Kolkovski, S., Hulata, G., Simon, Y., Segev, R. and Koren, A. (2003) Integrated Agri-Aquaculture Systems, A Resource Handbook for Australian Industry Development (eds. G.J. Gooley and F.M. Gavine). *Chapter 3 – Integration of agri-acuaculture systems – the Israeli experience*. Rural Industries Research and Development Corporation, RIRDC Publication, Kingston, ACT, Australia, pp. 14–23.

Krkosek M. (2005) from BBC news report, March 2005.

Langan R. (2008) The University of New Hamsphire's marine Aquaculture Center (Personal Communication).

New Zealand Geographic © (2004) Issue 90.

Olli, J.J. and Krogdahl, A. (2008) *Aquaculture Research*, **26** (11), 831–835.

Petersen, J.D., Rask, N., Bundgaard Madsen, H. et al. (2009) Eutrophication in coastal ecosystems. *Hydrobiologia*, **629**, 71–89.

Pimentel, D. (2006) in *Environment, Development and Sustainability*, vol. **18** (ed. S.S. Lang), Cornell University, pp. 119–137. Chronicle Online March 2006.

Poulin, B., Lefebvre, G., Allard, S. and Mathevet, R. (2009) Reed harvest and summer drawdown enhance bittern habitat in the Camargue. *Biological Conservation*, **142** (3), 689–695.

Pruginin, Y., Fishelson, L. and Koren, A. (1988) Proceedings from the 2nd International Symposium on Tilapia in Aquaculture, Bangkok.

Pullin, A.S. (1997) *Journal of Insect Conservation*, **1** (3), 177–185.

Rambo, A.T. (2007) in *Voices from the Forest* (ed. M. Carins), RFF Press, Washington, DC.

Rawlings, B. (2006) *The Great Salmon and Sea Trout Lochs of Scotland*, Swan Hill Press, Shrewsbury.

Rosauer, R. (ed.) (1987) adapted from Ancient Hawaiian Aquaculture, from the following: Barry Costa-Pierce, "Ancient Hawaiian Aquaculture" Bioscience May 87 Vol 37 No5 and from: Kikuchi, W.K., Science "Prehistoric Hawaiian Fishponds" July 76.

Smith, G.F., Bruce, D.G., Roach, E.B. *et al.* (2005) Assessment of recent habitat conditions of eastern oyster *Crassostrea virginica* bars in mesohaline Chesapeake Bay. *North American Journal of Fisheries Management*, **25** (4), 1569–1590.

Spencer, B., and Troy, P. (2009) Hawaii Ocean Technology. (Personal Communication).

Staples, C.R., Mattos, R., Risco, C.S. and Thatcher, W.W. (2009) Improving Cow Fertility Through Fishmeal Supplementation. University of Florida, IFAS Extension Document #D595.

Sutcliffe, J. (2010) Institute of Ecology and Environment Management. (Personal Communication).

The Seafish Organisation (2007) Plaice Factsheet 2, April 2007.

Walker, M. (ed.) (2009) Article – BBC Earth News, July 2009.

Webber, H.H. and Thurman, H.V. (1991) *Marine Biology*, 2nd edn, Harper Collins.

www.chesapeakebay.net (accessed January 2010).

www.cranfiled.ac.uk/sas/nsri (accessed May 2009).

www.defra.gov.uk/environment/quality/land/soil/sap/index.htm (accessed November 2009).

www.dragonfeeds.com.

www.gafta.com/fin (accessed June 2009).

www.lauguedoc-france.info (accessed June 2009).

www.lochduart.com

www.mangroveactionproject.org (accessed June 2009).

www.nzfreshwater.org (accessed June 2009).

www.sarnissa.org (accessed July 2009).

www.tourduvalat.org (accessed May 2009).

6 The future of farming and its implications for conservation

6.1 Introduction

In earlier chapters we have highlighted a number of case studies where farmers are making good progress in their conservation efforts, and our hope is that these examples will inspire others. We do, however, need to address some of the larger issues. As we look towards the future, farming has a seemingly impossible remit: To feed, clothe and supply energy to an ever increasing population using the same amount or less land, to do so; often using less natural resources such as water and fuel; to meet the requirements of ever more challenging legislation with respect to erosion and waste products, and so on; and to do all of this whilst encouraging and maintaining the presence and health of native wildlife.

In this final chapter we look at current farming practices and how they will need to adapt to increase productivity, using sustainable methods and new technologies, in light of current food security, and environmental concerns including biodiversity and global warming. Here we address the following topics:

- Industrial-scale farming and monocultures
- Biotechnology
- Implications of reduced crop and stock diversity
- Subsistence farming and emerging economies
  — Shade coffee growing
- Sustainable use of water, energy and waste:
- Agri-Environment schemes.

6.2 Industrial-scale farming and monocultures

For growers of food 'staples' such as rice, potatoes, wheat and vegetable oils, plus major fruit crops such as oranges and tomatoes, biofuel crops and of course grass for grazing stock; continuing mass production must be considered as an important way

Introduction to Wildlife Conservation in Farming   Edited by Stephen Burchett and Sarah Burchett
© 2011 John Wiley & Sons, Ltd

to meet consumer demands, but with mass production comes additional pressures both to the environment and for biodiversity. This industrial-scale production almost always involves high-tech mechanisation at all stages, from ground preparation, through seeding, spraying, watering, fertilisation to harvesting. Any consideration for natural ecological processes are largely put aside in favour of maximising yields, and the machinery used will be designed for a specific cropping system, resulting in vast monocultures. However, disease from soil-borne pathogens can spread rapidly through such systems if regular spraying with pesticides is not undertaken. Also, trees and hedges are often removed and topographical features sometimes flattened to accommodate the machinery – wildlife has little chance to get a foothold here.

It is possible to ameliorate some of the damaging effects of industrial-scale farming. For example, research by the University of Southampton, UK, in conjunction with the Game Conservancy Council, (www.gct.org.uk, accessed February 2010), has indicated the benefits of using 'Beetle Banks' on fields of 20 ha or more. These are strips or mounds of grass and perennial plants that supply a habitat to beneficial insects such as beetles and spiders that prey on pests such as aphids. Like field margins, discussed earlier, these can be economically utilised along edges of fields and difficult to cultivate areas of the land, and potentially results in a reduction in pesticide requirements. The beetle banks further supply a food source and nesting sites for birds such as the grey partridge (*Perdix perdix*). Their use has now spread internationally and Oregon State University, US, has collaborated with local farmers to establish many beetle banks, and has developed a website to help to describe the benefits and to facilitate their deployment (www.ipmnet.org, accessed March 2010).

Other possible methods include recycling of water and use of natural energy sources such as solar and wind power to reduce carbon footprint, or collaboration between crop and stock farmers, for example trading straw and lower grade crops for manure fertiliser. It is not clear just how many farming industrialists are taking on board such measures, what is clear is that there is a long way to go, and that education and legislation may be the only way forward if farmland wildlife is to have any future in such land.

In the US ecological studies indicate that watercourses on and downstream from farmland are often eutrophic and that the nutrients causing this eutrophication are between 50 and 70% derived from fertilisers, such nutrient enrichment of freshwater ecosystems leads to the destruction of all animal life in these water systems (Altieri, 2000). So, what hope for conservation on farmland such as this? Prophylactic applications of nutrients is not only bad news for aquatic wildlife, or indeed for the more far-reaching implications for drinking water as these nutrients penetrate the water table, but it actually does not make sound economic sense as they are clearly not benefiting the crop if they are being washed away. Educating the farmers on the science behind using more efficient applications, that is rates, timing and environmental conditions would decrease inputs, increase uptake and reduce waste. The important role of scientists in modern day farming and its associated obligations with respect to wildlife, resource and wildlife conservation cannot be overstated, from general data collection and dissemination, through plant breeding and tissue culture to recombinant DNA technology – a sophisticated, growing, if sometimes controversial, discipline.

6.3 Science and technology

Food supply is a multifunctional industry where crop production is governed by many factors including climate, biotic, economical, ethical, environmental and political issues and developments in science and technology.

Science and technology can help to address the constraints imposed on crop production systems by environmental factors but can simultaneously impose unforeseen constraints on adopting such scientific methods due to issues arising from public acceptance of these new technologies. However, the stark reality is the human race needs to produce 50% more food over the next 30–40 years (Royal Society, 2009) to meet increased demand from an ever-expanding human population. There are several ways forward that include expanding the area of land designated for crop production, improved varieties of crop plants, improved crop management systems and the integration of genetically modified (GM) organisms into agricultural systems. The first option is an easy short term fix but will fail to produce medium to long term gains in total crop yield, due to many factors, including the impact of climate change on contemporary cropping areas and yields (Royal Society, 2009).

Improving crop management systems is an ongoing process and in the developed world the focus is on the implementation of precision farming methods, whereas in subsistence communities and the emerging and transitional countries improved crop management systems may rely on the development of integrated pest management (IPM) systems.

6.4 Precision farming

Field scale food production systems in the developed world are increasingly relying on precision farming technology to enhance crop performance and yields while simultaneously reducing the impact of field operations on the environment. Such tools rely on sophisticated global positioning systems (GPSs) and computer-based technology, which is integrated into the cab of the tractor and the attached field implements such as the seed drill, sprayer and fertiliser spreader. Precision farming aims to control inputs into a crop system by mapping a number of variables that determine crop growth;

- Soil type and soil depth
- Nutrient status of soils
- Soil water status
- Weed and pest pressure.

These variables can be incorporated into the crop management plan and through integrated technologies crop performance can be managed at a very sophisticated level, precision of modern equipment has advanced from an initial accuracy of tens of metres to a few centimetres (www.precisionpays.com, accessed 7 March 2010) ensuring improved crop yield and reductions in variable costs and impacts of agrochemicals on the environment. The benefits to wildlife arise from mapping inputs to the exact requirements of the crop as it performs in the field. This approach removes many of the inaccuracies seen in the prescriptive field-based method; this is where a field crop is grown to a prepared nutrient and pesticide programme that is applied across the whole field. This is a blanket approach and is essentially wasteful. Precision farming results in applications of crop nutrients and pesticides where they are required, which in many cases is often in isolated patches within a field system such as a wet hollow, an exposed site, shallow and stony soil, or field patches prone to weed infestations. Targeting crop inputs reduces waste, improves field performance and prevents collateral damage from unsophisticated application of agrochemicals. Financial savings as well as environmental benefits can be made by restricting herbicide applications to known areas of weeds, restricting slug pellets to cloddy soils (establishing crops in cloddy soils exposes the young plants to heavy slug predation) thus reducing the levels of the active ingredient, Metaldehyde, in water catchments – which in Europe are governed by the Water Framework Directive.

Wide scale adoption of precision farming will help to secure food supplies by improving the capacity of growers to continue to deliver food in an ever demanding world market, enhance management options for crop pests and disease, in particular, help map the development of emerging resistant populations of pests and thus aid in the decision making process with respect to the selection of the active ingredient the pesticide.

6.5   Biotechnology

Advances in the performance of agricultural crop plants and livestock animals are historically rooted in biotechnology, by improving the yield and disease resistance of crop plants through traditional breeding programmes (Warren, Lawson and Belcher, 2008; Royal Society, 2009) and the development of our modern breeds of livestock (Hall and Clutton-Brock, 1989). These traditional methods of improving performance of crops and livestock animals are based on long term breeding programmes where desirable traits, such as disease resistance and drought tolerance in crop plants, and milk yield in dairy cattle, are established by crossing parents who express one or more of these desirable traits and selecting the progeny (off-spring) where these traits have recombined to produce a hybrid with improved agronomic characteristics.

Another traditional example is the development of modern apple varieties. In the UK the apple market has been transformed from an orchard system that was once dominated by standard trees (4.6–6 m in height) to contemporary orchards where trees are now either dwarf (up to 2 m in height) or semi-dwarfing (4–5 m in height). The conceptual element of this change is to simplify harvesting operations by removing the requirement for apple pickers to climb ladders. This advancement was achieved by grafting traditional desert apple scions from standard trees onto dwarfing rootstocks.

The final example of traditional crop breeding programmes is the development of the F1 hybrid. This breeding system develops progeny that have excellent field performance. Good examples come from the crops of field scale horticulture, such as the cauliflower (*Brassica oleracea* var. botrytis). Historically, cauliflowers are developed by the out-crossing breeding system, which results in progeny that head-up (development of the cauliflower pre-inflorescence – the harvestable component) in a variable pattern. This breeding system also results in plants that produce a very variable cauliflower with heads varying in size and appearance. These characteristics are undesirable for the end user; in particular the supermarkets, and therefore growers were fortunate if they could sell 70% of their crop (Fuller, personal communication, 1998), such a system is inherently wasteful. The development of the F1 breeding system has replaced these undesirable traits as the progeny of the parents have been selected for consistency in size and heading-up date, now growers can harvest 95% of their crop and sell through the supermarket system.

All the above examples are widely practiced agricultural and/or horticultural techniques and cause no undue concern in the market place. The traits bred into these new varieties have been completely manipulated by mankind and quite likely would never occur naturally. Traditional breeding programmes will continue to contribute to the future performance of agricultural systems but there is an urgent and pressing need to address a number of key issues that may not be able to be met by these traditional approaches. With respect to crop plants, particularly staples such as cereals and rice, the key environmental issues that need addressing are drought and salinity tolerance for cereals and for rice the incorporation of C4 photosynthetic mechanisms into the photosynthetic apparatus of the plant, which may help to improve yields.

The photosynthetic mechanism of rice is based on the C3 pathway, which is compromised in tropical ecosystems by a phenomenon known as photorespiration. This is where the initial CO_2 acceptor in the C3 pathway will either accept CO_2 or O_2, and in tropical ecosystems CO_2 becomes limiting during the heat of the day and consequently more O_2 is accepted resulting in further respiration (a normal process in plants following the initial fixation of CO_2) and consequently reduced yields. If C4 can be introduced into rice then yields could be increased by 50% as a result of removing yield-limiting photorespiration (Hibberd, Sheehy and Langdale, 2008) this would be significant progress and may help to alleviate food poverty for millions of people and potentially help to reduce the impact of subsistence farming on local ecosystems as improved yield need not mean expansion.

Cereal producing counties like Australia and the USA, and indeed the eastern counties of England, experience significant periods of prolonged water shortages or severe drought, where annual rainfall is between 300 and 800 mm (www.cgiar.com, accessed 8 March 2010). In other regions of the globe where agriculture is the mainstay of subsistence farmers, which include vulnerable regions such as Sub-Saharan Africa and Central Asia, drought is a major constraint to plant growth and crop performance (www.cgiar.com, accessed 8 March 2010). Contemporary technologies such as traditional plant breeding programmes, water harvesting, companion planting and seed mixtures (Lynch, 2007) have helped to improve crop performance in droughty areas but there is a limit to what can be achieved using these approaches. Advances in science offer agricultural science a range of new tools, which can help address pressing issues such as drought and other environmental stresses. These tools are based on molecular technology, some resulting in GM organisms using recombinant DNA technologies, others, such as marker technology, assisting in conventional plant breeding programmes.

In traditional plant breeding programmes the time consuming phase of the programme is progeny selection, which requires breeders to grow all the progeny lines and test individual off-spring for the expression of the desired trait. This step can take many years. But these desirable and defined traits are often confined to specific genes known as quantitative trait loci (QTL); these are stretches of DNA strongly associated with the gene for a specific trait (Royal Society, 2009). Molecular biologists have developed methods where they can insert a molecular DNA marker linked to the QTL. These markers can then be tested *in-vitro* thus improving the progeny selection process. Another important concept in plant breeding is that desirable traits, such as disease resistance and drought tolerance, are often associated with undesirable traits such as poor palatability and low yield, this is known as linkage drag. Plant breeders use molecular markers to help to identify rare plants in the breeding programme where desirable traits are not linked to undesirable traits, again speeding up the selection of improved varieties (Royal Society, 2009).

In 2007, 114.3 million ha were planted to GM crops, in 23 countries, including USA, Argentina, Brazil, Canada, India and China, and were grown by 12 million farmers (Wahlquist, 2008). Typically these crops were modified for herbicide resistance (Glyphosate resistance) and pest resistance using the bacterium *Bacillus thuringiensis* that produces a crystalline toxin (Bt toxin) that is specific to certain invertebrate groups, notably the caterpillars of butterflies and moths. Maize is a major food staple and has been modified by recombinant DNA technology to be both herbicide resistant and pest resistant, whereby expression of the Bt toxin is the mechanism of action for pest resistance. In the USA maize modified to express the Bt toxin is grown across the Corn Belt, where it is used to combat damage caused by the European corn borer (*Ostrinia nubilalis*), a major introduced pest species in the USA, which bores into the stem and cob of the maize plant. This technology, however, incurs non-target impacts on species diversity.

Concerns have been raised about the impact of Bt maize on the health and integrity of the charismatic monarch butterfly (*Danus plexippus*). In laboratory studies Losey, Rayor and Carter (1999) reported that larvae fed on Bt corn pollen and leaves of milkweed plants (*Asclepias spp.*) exposed to Bt corn pollen were susceptible to the Bt toxin. These findings raised concerns within the American Environmental Protection Agency (EPA) (Anon, 2004) because monarch butterflies are a protected species in the USA and about half their population frequent the US Corn Belt (Dively *et al.*, 2004). Furthermore pollen from Bt maize, can spread onto adjacent larval food plants, like milkweed, and this spurred the EPA to call for further research into the long term non-target effects of Bt maize on monarch butterfly populations. Results from this new round of research have concluded that the detrimental impacts of Bt toxin are minimal on the wider population of the monarch butterfly, with a conservative estimate of <0.8% of the population being exposed to the Bt toxin, giving estimates of infield mortality of around 23% (Dively *et al.*, 2004). The research team also concluded that there were no significant impacts on the sex ratio and wing lengths of these butterflies (Dively *et al.*, 2004). Debate and controversy still exists on the potential non-target impacts of Bt maize but the alternative method would be widespread use of insecticides (in the USA Pyrethrum based compounds are used, which are non selective and would kill the monarch caterpillars).

Another large-scale adoption of GM technology is herbicide resistance in crops like soya, maize and cotton that have been developed to be resistant to broad-spectrum herbicides like Glyphosate. These GM crops enable farmers to grow continuous rotations of a single crop and control numerous difficult weeds. However, this system is likely to break down as the resistance mechanism in the crop essentially results in a strong selection pressure on the in-field weed population. This selection pressure will, indeed is, resulting in the development of weed populations that have evolved resistance to Glyphosate (Anon, 2010).

Clearly genetic engineering has a key role to play in the future of food production systems but there are a number of major obstacles to overcome:

- Public acceptance of GM in Europe
- A requirement for detailed ecological studies on the impact of GM on the environment, ecosystems and trophic levels in food webs
- Detailed knowledge is required on the potential of GM crops to outcross with wild relatives
- Design and implementation of management strategies for field scale GM systems, to include the incorporation of buffer zones and refugia for wildlife
- The potential for GM crops to alleviate food poverty and improve the quality of life for subsistence farmers
- Development of multi-gene GM plants for crop systems in arid regions
  — In particular the potential for GM-induced drought resistant crops to yield viable and good quality grains in arid regions.

Globally there is a powerful impetus to raise small farmers from poverty associated with subsistence farming, science and technology is being implemented in a number of ways to help improve cropping systems so that these farmers can benefit from yield surpluses. One example of this is tissue culture.

6.6 Tissue culture

Tissue culture or micropropagation is a process where plants can be propagated vegetatively by excising small sections of donor plant material, normally from a meristematic node or apical region, and culturing these cuttings in sterile nutrient agar. The agar media contain all the essential elements for plant growth (Murashige and Skoog (1962) is a commonly used base media), which is modified by culture technicians to be specific to the target species and the requirements of the culture system. The influence tissue culture has had on plant breeding and the ornamental horticulture sector is significant, particularly with respect to the bedding plant industry and the rise in popularity of orchids. Orchid seeds have very limited starch reserves and in nature orchid seedlings rely on fungal associations to aid nutrient uptake, in the nursery sector this has been replaced by nutrient rich agar and standard tissue culture protocols. This approach has reduced the retail price of orchids considerably.

Tissue culture is widely used not only in the ornamental horticulture sector but also in research programmes associated with recombinant DNA technologies, and in plant conservation. Indeed endangered dipterocarps (Chapter 4) are being conserved in part by developments in tissue culture protocols (Nakamura, 2006). Micropropagation can also be used to eradicate long term disease in vegetatively propagated crops such as the banana crop (*Musa spp.*). The banana is an important food plant for many subsistence farmers in Eastern Africa (www.absfafrica.org, accessed 12 March 2010) however due to disease and pest pressure, yields have declined significantly in the last 10 years. One driver for the spread of disease, and consequently the decline in regional yields, was the practice of farmers sharing sucker propagules (bananas naturally propagate by clonal suckers), these propagules are frequently infected with surface pathogens from their parent plant and consequently the spread of numerous foliar pathogens is perpetuated. To break this cycle of disease, plant scientists required a source of disease-free material. To source such disease-free material, sucker plants were harvested and the apical meristem was excised from the surrounding parent material (leaves and leaf sheaths), surface sterilised (standard tissue culture practice) and aseptically cultured using micropropagation protocols. The resulting stock of banana plants was disease free and through an extension programme plants were distributed to 500,000 resource poor farmers across Kenya (www.absfafrica.org, accessed 12 March 2010). The results of this integrated approach has increased banana productivity from 20 to 45 tonnes/ha and household incomes from US$1 per day per family to US$3 per day per family. Success stories such as this need wider dissemination and further integration into subsistence agriculture and this can only be achieved through extension and education. The potential benefits of these technologies if implemented with care may help to reduce the devastating impacts of the bush meat trade. Clearly this last statement is very complex and requires considerable management of several cultural issues, but alleviating hunger and rural poverty in emerging and transitional economies is an essential and positive step in the right direction and science and technology (including GM) have a key role to play in the food security of people in poor rural communities.

6.7   Implications of reduced crop and stock diversity

Biodiversity has become almost a 'buzz word' for modern environmentalists, but it must not be forgotten that it is not just the wildlife that is under threat. Crop plants and stock animals are also affected by reduced diversity. Apples are a significant point in case. Britain is the natural home of the apple, with all varieties originating from the crab apple (*Malus sylvestris*). It is not really known just how many of these varieties still exist, but certainly there are many thousands. It is doubtful, however, that most people in Britain have tasted more than the half a dozen or so varieties that are available in the supermarkets, and often those that are sold have been imported from around the world. Apples that are available have been bred for yield, uniformity and disease resistance, though most of these are still heavily sprayed with pesticides. It is a similar situation with stock, particularly dairy cattle where the black and white Holstein cow predominates the dairy industry in temperate regions due mainly to its high yields. In Chapter 3 we indicated how older, hardier breeds are frequently used on smaller farms, as they are adapted to marginal areas. These are often the only source of genetic diversity in stock species, and this diversity is in the stewardship of breed societies and the Rare Breed Survival Trust, often with little financial support from governments.

One repercussion of the widespread uptake of developments in science and technology is the continued intensification of agriculture and clearly GM crops are a major component of this intensification. There is a real threat that GM crops can result in the narrowing of diversity in crop plants and hence overall biodiversity in crop genotypes. This narrowing in the genotypes is driven by more farmers growing an ever-decreasing range of crop plants to meet demands from the world market. The implications of this action have the potential to be disastrous for future generations if the original heirloom varieties are lost. The question is 'what role do GM crops play in driving the loss of crop diversity?' Probably no more than what has already occurred from the widespread cultivation of crop plants derived from traditional plant breeding programmes. Plants such as banana, rice, millet and wheat, were bred and selected for enhanced performance and have been widely distributed across the globe and have already reduced the diversity of food plants from 700 species to fewer than 200 species (Marinelli, 2004).

Undoubtedly there is a real threat to wild populations of plants as GM plants can outcross with wild relatives (Orton, 2003) changing the genetic diversity of wild populations. However, research in this area is an emerging issue and currently our knowledge of the potential effects of GM outcrossing is not fully known. Should this observation restrict the application of GM technology? If appropriate policy and safety protocols are implemented it seems unlikely that there is any greater risk to wild populations from GM crops than from crops modified by traditional plant breeding programmes.

The implications of reducing the range of food plants in agricultural systems and relying on simple monocultures for large-scale food production systems are quite significant with regards to food security. Furthermore monocultures reduce the integrity of the surrounding landscape matrix and the native animals that use this landscape (Chapter 2). To safeguard the future of genetic diversity in crop plants and more widely in native flora there is an impressive global effort to stem the tide in the illegal trade of endangered species, overseen by The Convention on International Trade in Endangered Species (CITES) and to deposit germplasm (seeds) into long term gene banks and seed collections, such as the Millennium Seed Bank (MSB), Kew (www.kew.org, accessed 12 March 2010). This is an ambitious project that aims to conserve 25% of the world's plant species by 2020 through a range of initiatives and global collaboration with universities and botanical gardens. One example of the impact of this work, coordinated by the MSB and in partnership with the Food and Agricultural Organisation (FAO), is the Difficult Seeds Project. This project aims to work with farmers and seed bank curators in Africa to identify, handle and conserve difficult seeds, in particular crop wild relatives, which are under-utilised species and hence are threatened by abandonment and neglect (www.kew.org, accessed 12 March 2010). These species may contain important genes for enhancing crop performance in the future where the cropping environment has been modified by climate change; two possible examples would be enhancing tolerance to drought and salinity.

6.8   Subsistence farming and emerging economies

Smaller farms have an important role to play with respect to crop and stock diversity, discussed above, but also to the wider biodiversity within the farming landscape. Many subsistence farmers in tropical regions are having a significant positive impact on biodiversity on their smallholdings. One such example is shade coffee growing.

6.8.1   Shade coffee

Originating in Ethiopia, coffee (*Coffea spp.*) is now grown in many tropical countries around the world. Coffee plants are forest understorey plants, however commercially grown coffee plants have been bred to withstand full sun. Sun-grown coffee plants are higher yielding and are grown as a monoculture in areas that were once forested; they are, of course, an important export commodity for many countries, however such plantations require considerable inputs of chemical fertilisers and pesticides, and are responsible for the removal of tropical rainforest, and its associated biodiversity, on a large scale. For example, in Central America 2.5 million acres (1 million ha) have been felled for coffee plantations (www.coffeehabitat.com – accessed January 2010).

Smaller farmers still grow their coffee in the shade of the forest trees, some using selective felling and partial removal of other understorey plants to reduce competition for nutrients, others growing their coffee 'rustically', that is planting amidst the prevailing flora on the forest floor. In Ethiopia many farmers grow coffee plants under individual shade trees as an addition to their farming activities, this has been shown to be enough to maintain a significant level of avian diversity (Gove *et al.*, 2008). In addition plant biodiversity benefits as these trees often harbour numerous epiphyte species (Hylander and Nemomissa, 2008). Preserving these singular trees, and growing a small crop in the shade of its branches, thus make a huge contribution to local biodiversity conservation. Though the yields of shade coffee may be lower, this method has a number of significant advantages; the tree canopy protects the plants from harmful solar radiation; the trees and other plants help improve the soil by providing leaf litter and stabilisation by their roots, (in contrast, erosion and subsequent run-off into watercourses is a major issue sun-grown coffee plantations). The trees also tap into minerals from deeper in the soil horizon increasing nutrient cycling, plus the presence of diverse plant life provides food and shelter for invertebrates and their predators, especially birds, such as the Abyssinian Woodpecker (*Dendropicos abyssinicus*) and the Yellow-Fronted Parrot (*Poicephalus flavifrons*) of Ethiopia; and the Painted Bunting (*Passerina ciris*) and the Wilsons Warbler (*Wilsonia pusilla*) in the tropical regions of North, Central and South America; all of which greatly reduces or negates the need for chemical inputs. For the farmer, this diverse environment also provides other commodity resources such as fruits and timber, especially relevant when coffee prices are depressed, (www.shadecoffee.org – accessed January 2010).

It is clear that, although sun-grown coffee provides higher yields, the benefits to the environment, and in helping to sustain rainforest biodiversity, of growing shade coffee, is now coming to the attention of conservation-minded consumers, especially in the United States where demand is increasing. This demand has given rise to a 'Shade-Coffee Certification' that attracts a premium price, and it has been postulated that this premium go direct to the producers to help to ensure the continuing will to conserve forest-sensitive species (Perfecto *et al.*, 2004).

6.9 Sustainable use of water, energy and waste

There are now many technologies aimed at sustainable use of resources, the literature and the Internet are replete with both good, and some more questionable examples. Many of these technologies are new, however, farmers and scientists are also taking a new look at older technology. Farming utilises considerable resources, plus farmed products need to be transported utilising yet more resources. Farming is therefore an obvious target market for such technologies. In Table 6.1 we summarise some of the systems, new, old-but with new perspective and integrated.

Table 6.1 Technologies utilised by the farming industry to facilitate sustainable practice

	Technology	Description
Alternative energy	Anaerobic digesters	Septic breakdown of waste materials in anaerobic conditions to produce biogas energy and compost the solid matter
	Biofuels	Any fuel made from biomass, for example large-scale fermentation of plant sugars from, for example maize and soya to make ethanol or biodiesel from plant oils or animal fats
	Co-firing	Alternative 'greener' fuels used to supplement fossils fuels and thus reduce fossil fuel emissions
	Micro-hydro	Small-scale hydro-electric power systems
	Solar power	Solar energy converter that generates electricity
	Wave power	Wave energy converter that generates electricity
	Wind turbines	Wind energy converter that generates electricity
Reduction of water use/loss	Micro-sprinkling/drip irrigation	Delivery of water direct to plants when and where required to reduce loss to the environment
	Retention reservoirs	Water catchment and storage structure to help reduce quantity of drawn water
Waste reduction	Precision agriculture	Computerised system delivering exact required inputs directly where needed thus reducing inputs and preventing excess wastes entering waterways and ground water
	Slurry injection	Precision injectors used to deliver slurry directly into the soil where and when required. In contrast, surface-spread slurry generates considerable waste and can be lost to run-off
Sustainable production	Aquaponics	Aquatic polyculture within an enclosed recirculating system
	Biological control	Use of predators to control pests
	Companion planting	Growing usually two (sometimes more) plant types close together to facilitate pest control and pollination, for example garlic grown near carrots to reduce carrot fly (*Psila rosae*) infestation

Table 6.1   (*continued*)

Technology	Description
Crop rotation	Growing different crops in rotation that utilise different soil horizons, soil nutrients and minerals. Also to help prevent pests and diseases from being carried over to a new generation of the same crop. Fallow years are often used in this system
Green manure crops	High nutrient crop grown between harvest and resowing. Helps to suppress weeds, reduce water loss, and when mature is not harvested, but ploughed back in to improve soil fertility, for example alfalfa (*Medicago sativa*)
Ground cover	Cover crop to out-compete weed incursion and help reduce water loss from soil
Hydroponics	Plant production in water and nutrient solution in enclosed, controlled system reducing the likelihood of pathogen incursion
Integrated pest management (IPM)	Combination of pest management strategies, both natural and chemical, designed to maximise efficiency whilst reducing chemical inputs
Polyculture	System with two or more crops or stock that are inter-dependent resulting in reduced inputs

The pressing need to conserve water is inevitably going to impact on crop systems and again this threatens food security. Crops that are susceptible to drought, which include staples such as wheat and barley, can be manipulated by traditional plant breeding programmes or by biotechnology, but there is a limit to enhancing drought tolerance. This implies the need for other innovations such as improved irrigation systems, integrated crop management and remote sensing of soil water status. Integrated crop management systems rely heavily on companion planting and intercropping, such as the shade coffee example and the push-pull me approach adopted in Kenya for maize production.

Maize is an important crop in East Africa and has a major role in securing food supplies and cash income for many Kenyan farmers. The push-pull me approach is an IPM strategy where maize is grown with two other plant species in order to reduce the impact of an invertebrate pest, the maize stalk borer (*Busseola fusca*) and spotted stem borer (*Chilo partellus*) which together can reduce yield by as much as 40%, and the parasitic weed known as witchweed (*Striga hermonthica*) which can reduce yield by another 30–50%, these pests impact 40% of Africa's arable land (Royal Society, 2009). The basic principle of push-pull me in the maize field is surrounded by a border of forage grass, napier grass (*Pennisetum purpreum*) which is more attractive to the insect pests (moths) than the maize, and thus the moths lay the damaging larvae into the stems of the napier grass. Furthermore the napier grass produces a gum-like substance that kills the pest, thus reducing the local pest population, this element is known as the 'pull'. The 'push' component of this integrated approach is the sowing of *Desmodium uncinatum*, a forage legume, (sometimes called Spanish Tick-Clover but known by a number of other common names), in between the rows of maize. The *Desmodium* produces semiochemicals that repel the stem borer moths from the maize (the push). An unforeseen advantage of the *Desmodium* is it produces root exudates that are toxic to the *Striga*, and finally the ground cover provided by *Desmodium* helps to conserve soil moisture (Royal Society, 2009).

Cultural methods used to improve soil water status rely on good soil structure, achieved by reducing damaging practices such as soil compaction. Adopting good farming practice, and enhancing the reserves of soil carbon and soil organic matter can accomplish this. This requires inputs of farmyard manure (FYM) and/or an adoption of green manure crops (Chapter 2).

Soils in cereal production areas around the globe are classified as either degraded or very degraded, and if arable systems in these areas are going to improve the organic matter content of their soils, they would need to consider adopting a mixed farming model where livestock are incorporated into the farming system. This may rely on housing winter stock, and feeding stock with grain, the volume of grain-fed livestock can be reduced by adopting a sustainable stocking density, using traditional breeds and incorporating traditional fodder crops into the arable rotation, but also the benefits to soil organic matter will reduce the loss of soil moisture and consequently help to reduce the volume of irrigation required. The aim of this approach would be to reduce the water footprint (WF) of agriculture.

Irrigated crops account for 50% of the world's food supply (Royal Society, 2009) and in many cases irrigation occurs in arid and fragile regions of the globe, but in countries where water is not essentially limiting (the UK for example) then designing farming systems that reduce the volume of irrigation water will help to improve the WF of agriculture. The WF of UK agriculture is 73% of the nation's total WF (Royal Society, 2009).

6.10   Agri-environment schemes

Across Europe, the US and other developed countries, agriculture is directly affected by government policy that can drive producers in two major directions. The most obvious direction is the intensification of production and the enhancement of yield. Clearly this was the aim of Common Agricultural Policy (CAP) before the CAP reforms of 2003. Since 2003 farmers in Europe have been encouraged to produce food using practices that mitigate against environmental damage, these reforms have driven the producers in the opposite direction to their pre-2003 aims – the extensification of agriculture – where producers are reducing pressure on ecosystems by reducing inputs (Case Study 3.3).

The future for agri-environment schemes is unclear but agricultural support in developed countries will most likely continue in some form, as this approach for support payments is politically attractive and such schemes align with the Organisation for Economic Co-operation and Development (OECD) traffic light rules and find political acceptance under World Trade Organisation (WTO) rules (Warren, Lawson and Belcher, 2008). The level of payments to farmers is most likely to be more closely associated with measurable deliverables (i.e. an increase in species diversity and abundance) and indeed such practice is partially in place in the UK with the implementation of two tier agri-environment schemes (the Entry Level Scheme (ELS) and HLES scheme) where HLES schemes are targeted to areas of immediate conservation concern or of high conservation value. Warren, Lawson and Belcher (2008) give a more detailed discussion on the future of agri-environment policy, but what is not clear is the future impact of issues associated with food security. If policy is only focused on environmental issues then there could be considerable restriction in the development of new agricultural technology and the ability of agriculture to feed an ever-increasing population. There is no doubt the only way forward is to increase public awareness of agriculture, improve dissemination of good agricultural practice, widely achieved in the US through the extension programme coordinated by the United States Department of Agriculture (USDA). No such scheme exists in the UK and farmers rely on non-government organisations such as the Farming and Wildlife Advisory Group (FWAG). Some advice is available through the Department for Environment, Food and Rural Affair (DEFRA) and the Environment Agency, but this is associated more with compliance to regulations.

One positive development in agricultural policy in the UK would be the widening of public participation and funding for the development of educational institutions and research facilities. This measure would support future developments in agriculture and help to secure food supplies. Educational and research facilities have been significantly eroded in the UK by successive governments, who have adopted the political view that food security in the UK is not an issue, tantamount to burying the head in the sand approach, of course ostriches don't bury their heads in sand, but governments do fail in their responsibility to support long term national policy that would help to harmonise agriculture and conservation, and this has clearly weakened UK agriculture, a sentiment supported by a recent Royal Society report (Reaping the Benefits, Royal Society, 2009).

An anecdotal example that supports the above is seen in how the education of rural issues is being conducted in UK schools. UK farmers are accommodating school trips, but the majority of these trips are for primary school children (5–11 years of age) and while this is a good start there is an appalling lack of engagement with secondary school children (11–16 years of age). This is a poor situation resulting in exceptionally poor understanding of food production systems and often-biased opinions of agriculture, consequently a section of society unable to engage in informed debate about farming issues. This situation is further handicapping students entering conservation degree programmes at universities. Entry-level students have little or no knowledge of such rural issues and have selected their degree based on a polarised view of conservation. Future policy should aim to address this issue and strike an appropriate balance between production systems, conservation and education, failing to achieve this will further erode the ability of agriculture to feed people and mitigate against environmental damage.

Indeed this philosophical point should be the cornerstone of rural policy across the globe and it is organisations like the ITTO which have the political will to engage in this area, which have helped to improve awareness and understanding of tropical forest ecosystems, and consequently have contributed to improved forestry practice. The situation is not perfect but without International Tropical Timber Organisation (ITTO) support deforestation and degradation would be significantly worse than what has been observed in the twentieth century.

6.11   Conclusion

It becomes apparent that, whereas there is clear evidence that farming practices generally have had detrimental effects on the wildlife and the global environment, farmers who have reservations about their custodial role would do well to look to their more enlightened fellows to see that farming and conservation can, and do, go hand in hand. It should also be noted that in the so called 'western world' we are inclined to dismiss local knowledge and lore held by people who have farmed their regions for many generations – we do so at our peril. If we base our current understanding on a combination of traditional knowledge, recognition of environmental changes, modern scientific research and profound desire to ensure our planet survives intact, only then do we have a full compliment of the available tools, data and impetus with which to go forward.

References

Altieri, M.A. (Revised 2000) *Modern Agriculture: Ecological Impacts and the Possibilities for Truly Sustainable Farming*, University of California, Berkeley. Available at www.cnr.berkeley.edu (accessed March 2010).

Anon (2004) Investigating the long-term effects of Bt maize. Monarch butterflies: A threat to individual caterpillars, but not to the population as a whole. *GM Safety*, Available at http://www.GM-safety.eu/en/archive/2004/314.docu.html (accessed 7 March 2010).

Anon (2010) USA: Superweeds encouraged by GM Plants? Available at http://www.GM-safety.eu/en/news/731.docu.html (accessed 8 March 2010).

Dively, G.P., Rose, R., Sears, M.K. *et al.* (2004) Effects on monarch butterfly larvae (Lepidotera: Danaidae) after continuous exposure to CrylAb-expressing corn during anthesis. *Environmental Entomology*, **33** (4), 1116–1125. Available at http://www.GM-safety.eu/pdf/dokumente/bt-monarch-maryland.pdf, (accessed 8 March 2010).

Gove, A.D., Hylander, K., Nemomissa, S. and Shimelis, A. (2008) Ethiopian coffee cultivation – Implications for bird conservation and environmental certification. *Conservation Letters*, **1**, 208–216.

Hall, S.J.G. and Clutton-Brock, J. (1989) *Two Hundred Years of British Farm Livestock*, British Museum.

Hibberd, J.M., Sheehy, J.E. and Langdale, J.A. (2008) Using C-4 photosynthesis to increase the yield of rice – rationale and feasibility. *Current Options in Plant Biology*, **11**, 228–231.

Hylander, K. and Nemomissa, S. (2008) Home garden coffee as a repository of epiphyte biodiversity in Ethiopia. *Frontiers in Ecology and the Environment*, **6** (10), 524–528.

Losey, J.E., Rayor, L.S. and Carter, M.E. (1999) Transgenic pollen harms monarch larvae. *Nature*, **399**, 214.

Lynch, J.P. (2007). Roots of the second green revolution. Australian Journal of Botany 55, 493-512.

Marinelli, J. (2004) *Plant*, Dorling and Kindersley, London.

Murashige, T. and Skoog, F. (1962). A revised medium for rapid growth and bioassays with tobacco tissue cultures. Physiologia. Plantarum. 15, 472-497.

Nakamura, K. (2006) Micropropagation of *Shorea roxburghii* and *Gmelina arborea* by shoot apex culture, in *Plantation Technology in Tropical Forest Science* (eds K. Suzuki, K Ishii and S. Sakurai), Springer, Tokyo.

Orton, L. (2003) GM crops, going against the grain. Available at http://www.actionaid.org/docs/gm_against_grain.pdf (accessed 12 March 2010).

Perfecto, I., Vandermeer, J., Mas, A. and Soto Pinto, L. (2004) Biodiversity, yield and shade coffee certification. *Ecological Economics*, **54** (4), 435–446.

Royal Society (2009) *Reaping the Benefits. Science and Sustainable Intensification of Global Agriculture*, The Royal Society.

Wahlquist, A. (2008) Support Grows for GM. Available at http://www.hexima.com.au/pdf_files/articles/Support_grows_for_GM-Australian.pdf (accessed 8 March 2010).

Warren, J., Lawson, C. and Belcher, K. (2008) *The Agri-Environment*, Cambridge University Press, Cambridge.

www.kew.org. Millennium Seed Bank. Available at www.kew.org/science-conservation/conservation-climate-change/millennium-seed-bank/index.htm (accessed 12 March 2010).

www.kew.org. Difficult Seeds Project. Available at www.kew.org/science-conservation/conservation-climate-change/millennium-seed-bank/projects-partners/more-seed-projects/difficult-seeds-project/ (accessed 12 March 2010).

www.cgiar.org. Drought tolerant Crops for Drylands. Available at www.cgiar.org/impact/global/des_fact2.html (accessed 12 March 2010).

Acronym list

ACP	African, Caribbean and Pacific Countries	–
AEWA	African-Eurasian Water Agreement	–
AONB	Areas of Outstanding Natural Beauty	UK
APF	American Prairie Foundation	US
APFORGEN	Asia Pacific Forest Genetic Resources Programme	Asia
BAP	Biodiversity Action Plan	Europe
BCC	Biodiversity Conservation Centre	Russia
BOD	Biological Oxygen Demand	–
BYDV	Barley Yellow Dwarf Virus	–
CAP	Common Agricultural Policy	Europe
CCRP	Continuous Conservation Reserve Program	US
CDTF	Community Development Trust Fund	Africa
CRP	Conservation Reserve Program	US
CSS	Countryside Stewardship	UK
dbh	Diameter at Brest Height	Global
DEFRA	Department for Environment, Food and Rural Affairs	UK
DO	Dissolved Oxygen	–
DPS	Dartmoor Pony Society	UK
DVCA	Danum Valley Conservation Area	Malaysia
EEC	European Economic Community	Europe
ELS	Entry Level Scheme	UK
EPA	Economic Partnership Agreements	
EPA	Environmental Protection Agency	US
EPBC	Environmental Protection and Biodiversity Conservation	Australia
ETE	Emerging and Transitional Economies	–
FACE	Forest Absorbing Carbon Dioxide Emissions	Netherlands
FAO	Food & Agricultural Organisation	Global
FC	Forestry Commission	UK
FCS	Forestry Commission Scotland	UK
FCS	Food Certification Scotland	UK
FOGRIS	Forest Genetic Resources Information System	Malaysia
FRIM	Forest Research Institute Malaysia	Malaysia

FRS	Fisheries Research Services	UK
FSC	Forest Stewardship Council	UK
FWAG	Farming and Wildlife Advisory Group	UK
FYM	Farmyard Manure	–
GAI	Green Area Index	UK
HGCA	Home Grown Cereal Authority	UK
HLS	Higher Level Scheme	UK
IIP	Investors in People	–
IPM	Integrated Pest Management	–
IPU	Isoproturon	–
ITTO	International Tropical Timber Organisation	Global
IUCN	International Union for Conservation of Nature	Global
LaM	Lever and Mulch	Scotland
LFS	Less Favoured Area	Europe
LLP	Long Leaf Pine	US
LPO	Protection des Oiseaux	France
MSB	Millennium Seed Bank	UK
MUS	Malaysian Uniform System	Malaysia
NFA	National Forest Act	Malaysia
NGO	Non-Governmental Organisations	–
NH	New Hampshire	US
NIFO	Non-Industrial Forest Owners	–
NNR	National Nature Reserve	–
NPWS	National Parks and Wildlife Service	UK
NRCS	Natural Resources Conservation Service	US
NRM	National Resource Management	Australia
NRSI	National Soil Resources Institute	UK
NS	Nationally Scarce	–
NSA	National Scenic Area	–
NT	National Trust	UK
NVC	National Vegetation Classification	–
NVZ	Nitrogen Vulnerable Zone	–
OECD	Organisation for Economic Co-operation and Development	–
OELS	Organic Entry Level Scheme	UK
PCB	Polychlorinated Biphenyls	–
PF	Planted Forest	–
PFE	Permanent Forest Estate	Malaysia
PPE	Personal Protective Equipment	–
RBST	Rare Breeds Survival Trust	UK
RDB	Red Data Book	Global
RSPB	Royal Society for the Protection of Birds	UK
RSPCA	Royal Society for the Prevention of Cruelty to Animals	UK

SCS	Soil Conservation Service	US
SEPA	Scottish Environmental Protection Agency	UK
SERAP	South East Asia Research Programme	SE Asia
SEZ	Seedling Exclusion Zone	–
SFM	Sustainable Forestry Management	UK
SFP	Single Farm Payments	UK
SINC	Site of Importance for Nature Conservation	UK
SMS	Selective Management System	Malaysia
SNH	Scottish Natural Heritage	UK
SNS	Soil Nitrogen Supply	–
SOI	Sunart Oakwoods Initiative	UK
SPA	Special Protection Area	–
SRC	Short Rotation Coppice	–
SSSI	Site of Special Scientific Interest	Europe
STEEP	Solutions to Environmental and Economical Problems	US
TRC	Triangle Land Conservancy	US
UELS	Uplands Entry Level Scheme	UK
USDA	United States Department of Agriculture	US
USFWS	United States Fish & Wildlife Service	US
USGS	United States Geological Service	US
VOC	Volatile Organic Compounds	–
WES	Wildlife Enhancement Scheme	UK
WFD	Water Framework Directive	Europe
WRP	Wetlands Reserve Programme	US
WTO	World Trade Organisation	Global
WWF	World Wildlife Fund	Global

Species tables

Common Name	Scientific Name
Flora	
African mahogany	*Khaya ivorensis*
Alder	*Alnus glutinosa*
Alfalfa	*Medicago sativa*
Annual meadow	*Poa annua*
Ash	*Fraxinus spp*
Aspen	*Populus tremula*
Autumn squill	*Scilla autmnalis*
Banana	*Musa spp*
Barley	*Hordeum vulgare*
Bastard toadflax	*Thesium humifusum*
Bee orchid	*Ophrys apifera*
Beech	*Fagus sylvatica*
Beyrich threeawn	*Aristida beyrichiana*
Birch	*Betula spp*
Bird cherry	*Prunus padus*
Birds nest fern	*Asplenium nidus*
Black grass	*Alopecurus myosuroides*
Black medic	*Medicago lupulina*
Blueberry	*Vaccinium spp*
Bluegrass	*Poa spp*
Bluestem grasses	*Schizachyrium scoparium & Andropogon spp*
Bracken	*Pteridium aquilinum*
Bramble	*Rubus fruticosus*
Broom	*Genisteae spp*
Bugle	*Ajuga reptans*
Camphor	*Dryobalanops aromatica*
Carline thistle	*Carlina vulgaris*
Cauliflower	*Brassica oleracea*
Celandine	*Ranunculus ficaria*

Common Name	Scientific Name
Chickweed	*Stellaria media*
Cleaver	*Galium aparine*
Clover	*Trifolium spp*
Coffee	*Coffea spp*
Common chickweed	*Stellaria media*
Common field speedwell	*Veronica persica*
Common poppy	*Papaver rhoeas*
Common thistle	*Cirsium vulgare*
Corn marigold	*Chrysanthemum segatum*
Corncockle	*Agrostemma githago*
Cornflower	*Centaurea cyanus*
Cotoneaster	*Cotoneaster spp*
Cotton	*Gossypium spp*
Cowslip	*Primula veris*
Crab apple	*Malus sylvestris*
Creeping thistle	*Cirsium arvense*
Devil's bit scabious	*Succisa pratensis*
Docks	*Rumex spp*
Dog rose	*Rosa canina*
Dog violet	*Viola riviniana*
Douglas fir	*Pseudotsuga menziesii*
Downy birch	*Betula pubescens*
Downy brome	*Bromus tectorum*
Dulse	*Palmeria palmate*
Dwarf spurge	*Euphorbia exigua*
Eelgrass	*Zostera marina*
Einkorn wheat	*Triticum monococcum*
Elm	*Ulmus spp*
Emmer wheat	*Triticum dicoccum*
English bluebell	*Hyacinthoides non-scriptus*
English oak	*Quercus robur*
European gorse	*Ulex europaeus*
Eyebright	*Euphrasia spp*
Fescue grasses	*Festuca spp*
Field brome	*Bromus arvensis*
Field forget-me-not	*Myosotis arvensis*
Field pansy	*Viola arvensis*
Flax	*Linum spp*
Fodder radish	*Beta vulgaris*
Forage legume	*Desmodium uncinatum*

Common Name	Scientific Name
Frog orchid	*Coeloglossum viride*
Fumitory	*Fumaria spp*
Greater knapweed	*Centaurea scabiosa*
Groundsel	*Senecio vulgaris*
Guelder rose	*Viburnum opulus*
Hairy leafed apitong	*Dipterocarpus alatus*
Hairy vetch	*Vicia villosa*
Harebell	*Campanula rotundifolia*
Hawkbit	*Leontodon spp*
Hawthorn	*Crataegus monogyna*
Hazel	*Corylus avellana*
Hickory	*Carya spp*
Holly	*Ilex aquifolium*
Honeysuckle	*Lonicera periclymenum*
Indigo	*Baptisia spp*
Ivy	*Hedera helix*
Ivy-leaved speedwell	*Veronica hederifolia*
Japanese knotweed	*Polygonum cuspidatum*
Knapweed	*Centaurea spp*
Lady orchid	*Orchis purpurea*
Larch	*Larix decidua*
Late spider orchid	*Ophrys fuciflora*
Lime	*Tilia spp*
Lobarion lichen	*Lobarion pulmonariae*
Loblolly pine	*Pinus taeda*
Longleaf pine	*Pinus palustris*
Lupins	*Lupinus angustifolia*
Maize	*Zea mays*
Mat grass	*Nardus stricta*
Mayweeds	*Tripleurospermum spp,* *Matricaria spp,* *Anthemis spp,* *Chamaemelum spp*
Meadow clary	*Salvia pratensis*
Melapi pa'ang	*Shorea bracteolate*
Mengaris tree	*Koompassia excelsa*
Meranti merah	*Shorea singkawang*
Meranti species	*Shorea roxburghii* *Shorea pachyphylla*
Milkweed	*Asclepias spp*

Common Name	Scientific Name
Monkey orchid	*Orchis simia*
Mulberry	*Morus spp*
Mullien	*Verbascum spp*
Napier grass	*Pennisetum purpreum*
Nettle	*Urtica spp*
Night flowering catchfly	*Silene noctiflora*
Not available	*Dipterocarpus intricatus*
Oak	*Quercus spp*
Oat	*Avena sativa*
Oil palm	*Elaeis guineensis*
Pea	*Pisum elatiu*
Pedunculate oak	*Quercus robur*
Pennisetum grasses	*Pennisetum spp*
Perennial ryegrass	*Lolium perenne*
Perennial sowthistle	*Sonchus arvensis*
Phacelia	*Phacelia tanacetifolia*
Polypody fern	*Polypodium vulgare*
Purple moor grass	*Molinia caerulea*
Rabbit foot clover	*Trifolium arvense*
Ragged Robin	*Lychnis flos-cuculi*
Ragwort	*Senecio jacobaea*
Red dead nettle	*Lamium purpureum*
Red maple	*Acer rubrum*
Red oak	*Quercus rubra*
Reed sweet grass	*Glyceria maxima*
Resak labuan	*Vatica umbonata*
Rhododendron	*Rhododendron ponticum*
Rosebay willowherb	*Epilobium. angustifolium*
Rough poppy	*Papaver hybridum*
Rough stalked meadow grass	*Poa trivialis*
Round-leaved fluellen	*Kickxia spuria*
Rowan	*Sorbus aucuparia*
Sagebrush	*Artemisia tridentate*
Sainfoin	*Onobrychis viciifolia*
Sallow	*Salix cinerea*
Sand pine	*Pinus clausa*
Sandhill lupine	*Lupinus perennis*
Seagrass	*Thalassia spp*
Sedaman	*Macaranga triloba*
Seraya kepong	*Shorae ovalis*

Common Name	Scientific Name
Seraya kuning runcing	*Shorea acuminata*
Seraya punai	*Shorea parvifolia*
Sessile oak	*Quercus petraea*
Seyra tembaga	*Shorea leprosula*
Shepherd's needle	*Scandix pectin-veneris*
Shore dock	*Rumex rupestris*
Sitka spruce	*Picea sitchensis*
Slash pine	*Pinus elliotti*
Small cow wheat	*Melampyrum sylvaticum*
Small flowered catchfly	*Silene gallica*
Southern magnolia	*Magnolia grandiflora*
Speedwell	*Veronica persica*
Spring gentian	*Gentiana verna*
Starfruit	*Damasonium alisma*
Stinking chamomile	*Anthemis cotula*
Stinking hawksbeard	*Crepis foetida*
Stitchwort	*Stellaria holostea*
Strangler fig	*Ficus spp*
Sugar kelp	*Laminaria saccharina*
Sugar maple	*Acer saccharum*
Sun spurge	*E helioscopia*
Sunflower	*Helianthus annuus*
Sweet birch	*Betula lenta*
Sweet chestnut	*Castanea sativa*
Sweet gum	*Liquidambar styraciflua*
Switchgrass	*Panicum virgatum*
Sycamore	*Acer pseudoplatanus*
Telinga gajah	*Macaranga gigantifolia*
Turkestan alyssum	*Alyssum turkestanium*
Two-row barley	*Hordeum distichon*
Urat mata daun licin	*Parashorea malaanonan*
Venus's-looking glass	*Legousia hybrida*
Vetch	*Vicia spp*
Violet	*Viola riviniana*
Viper's bugloss	*Echium vulgare*
Walnut	*Juglans spp*
Western gorse	*Ulex gallii*
Wheat	*Triticum spp*
White birch	*Betula papyrifera*
White pine	*Pinus strobes*

Common Name	Scientific Name
White thingan	*Hope odorata*
Whorled caraway	*Carum certiculatum*
Wicthweed	*Striga hermonthica*
Wild daffodil	*Narcissus psudonarcissus*
Wild mint	*Thymus spp*
Willow	*Salix spp*
Winter barley	*Hordeum vulgare*
Winter rape	*Brassica napus*
Winter rye	*Secale cereale*
Winter wheat	*Triticum aestivum*
Wiregrass	*Aristida stricta*
Wych elm	*Ulmus glabra*
Yarrow	*Achilla spp*
Yew	*Taxus baccata*

Fungi

Morels	*Morchella spp*
Shiitake mushrooms	*Lentinula edodes*

Invertebrates

Atlantic oyster	*Crassostrea virginica*
Black urchin	*Paracentrotus lividus*
Bloodworm	*Glycera spp*
Blue crab	*Callinectes sapidus*
Caddis fly	*Trichoptera spp*
Carrot fly	*Psila rosae*
Chalkhill blue butterfly	*Polymmatu coridon*
Cuckoo-bee	*Nomada sexfasciata*
Digger wasp	*Aphilanthops spp*
European corn borer	*Ostrinia nubilalis*
Golden sun moth	*Synemon plana*
High brown fritillary	*Argynnis adippe*
Large copper butterfly	*Lycaena dispar*
Long-horned bee	*Eucera longicornis*
Maize stalk borer	*Busseola fusca*
Marsh fritillary butterfly	*Euphydryas aurinia*
Millipede	*Cylindriulus punctatus*
Mining bee	*Andrena spp*
Monarch butterfly	*Danus plexippus*
New Zealand crayfish	*Paranephrops planifrons*
Pearl-bordered fritillary butterfly	*Boloria euphrosyne*

Common Name	Scientific Name
Polychaete worm	*Neiris virens*
Sea lice	*Lepeophtheirus salmonis*
Sea squirt	*Tunicata spp*
Singapore freshwater crab	*Johura singaporaensis*
Spiny rock lobster	*Jasus edwardsii*
Sponge	*Porifera spp*
Spotted stem borer	*Chilo partellus*
White admiral	*Limenitis camilla*

Reptiles
Prairie king snake	*Lampropeltis calligaster*
Prairie rattle snake	*Crotalus viridis*
Sriped legless lizard	*Delma impar*
Western hognose snake	*Heterodon nasicaus*

Amphibians
African clawed frog	*Xenopus laevis*
Blandings turtle	*Emydoidea blandingii*
Blue spotted salamander	*Ambystoma laterale*
Marbled salamander	*Ambystoma opacum*
Natterjack toad	*Epidalea calmita*
Northern cricket frog	*Acris crepitans*
Spotted salamander	*Ambystoma maculatum*
Wood frog	*Rana sylvatica*

Fish
Atlantic cod	*Gadus morhua*
Atlantic salmon	*Salmo salar*
Atlantic sea bass	*Dicentrarchus labrax*
Bigeye tuna	*Thunnus obesus*
Longfin eel	*Anguilla dieffenbachia*
Rainbow trout	*Oncorhynchus mykiss*
Roach	*Rutilus rutilus*
Sea trout/Brown trout	*Salmo trutta*
Snapper	*Lutjanus spp*
Yellowfin tuna	*Thunnus albacares*

Birds
Abyssinian woodpecker	*Dendropicos abyssinicus*
Asian pied hornbill	*Antbracocerus albirostris*
Barn owl	*Tyto alba*

Common Name	Scientific Name
Bean goose	*Anser fabalis*
Bewick swan	*Cygnus columbianus*
Black grouse	*Tetrao tetrix*
Black Redstart	*Phoenicurus ochruros*
Bob white quail	*Colinus virginianus*
Bobolink	*Dolichonyx orzorus*
Brewers sparrow	*Spizella breweri*
Capercaillie	*Tetrao urogallus*
Cirl bunting	*Emberiza cirlus*
Corn bunting	*Miliaria calandra*
Crossbill	*Loxia curvirostra*
Curlew	*Numenius arquata*
Eastern bluebird	*Sialia sialis*
European bittern	*Botaurus stellaris*
Finches	*Fringilla and Carduelis spp*
Firecest	*Regulus ignicapillus*
Goldcrest	*Regulus regulus*
Golden plover	*Pluvialis apricaria*
Grasshopper sparrow	*Ammodrammus savannarum*
Grasshopper warbler	*Locustella naevia*
Great blue heron	*Ardea herodias*
Greater prairie chicken	*Tympanicus cupidus*
Grey partridge	*Perdix perdix*
Hawfinch	*Coccothraustes coccothraustes*
Hobby	*Falco subbuteo*
Hooded merganser	*Lophodytes cucullatus*
Hoopoe	*Upupa epops*
Kiwi	*Apteryx spp*
Lapwing	*Vanellus vanellus*
Little owl	*Athene noctua*
Meadow pipit	*Anthus pratensis*
Nightingale	*Luscinia megarhynchos*
Nightjar	*Caprimulgus europaeus*
Painted bunting	*Passerina ciris*
Plains wanderer	*Pedionomus torquatus*
Ptarmigan	*Lagopus mutus*
Red cockaded woodpecker	*Picoides boralis*
Red grouse	*Lagopus lagopus*
Redstart	*Phoenicurus phoenicurus*
Reed bunting	*Emberiza schoeniclus*

Common Name	Scientific Name
Rhinoceros hornbill	*Buceros rhinoceros*
Sage grouse	*Centrocerus urophasianus*
Sage sparrow	*Amphispiza belli*
Sage thrasher	*Oreoscoptes montanus*
Savanna sparrow	*Passerculus sandwichensis*
Skylark	*Alauda arvensis*
Spotted flycatcher	*Musciapa striata*
Stone curlew	*Burhinus oedicnemus*
Swallow	*Hirundo rustica*
Swift	*Apus apus*
Thrush	*Turdus spp*
Whip-poor-will	*Caprimulgus vociferous*
Whitethroat	*Sylvia communis*
Wigeon	*Anas penelope*
Wild turkey	*Melegris gallopavo*
Wilsons warbler	*Wilsonia pusilla*
Wood duck	*Aix sponsa*
Woodcock	*Scolopax minor*
Woodlark	*Lullula arborea*
Yellow-fronted parrot	*Poicephalus flavifrons*
Yellowhammer	*Emberiza citrinella*

Mammals

Common Name	Scientific Name
Beaver	*Castor canadensis*
Civet cat	*Hemigalus spp*
Clouded leopard	*Neofelis diardi subsp borneensis*
Coyote	*Canis latrans*
Dormouse	*Muscardinus avellanarius*
Greater horseshoe bat	*Rhinolophus ferrumequinum*
Hare	*Lepus europaeus*
Long eared bat	*Plecotus auritus*
Orang utan	*Pongo pygmaeus*
Pipistrelle bat	*Pipistrellus pipistrellus*
Plains pocket gopher	*Geomys bursarius*
Pronghorn	*Antilocapra americana*
Pygmy rabbit	*Brachylagus idahoensis*
Red squirrel	*Sciurus vulgaris*
Sagebrush Vole	*Lemmiscus curtatus*

Glossary

Biodiversity	Variation in life forms across the globe.
Biological Oxygen Demand (BOD)	The amount of oxygen taken up by microbes from decomposing organic matter in water.
C3 Species	Photosynthetic pathway in plant species where carbon dioxide combines with ribulose bisphosphate. Ribulose bisphosphate can accept both carbon dioxide and oxygen and photosynthesis can be rate limited in the tropics.
C4 Species	Where the initial fixation of carbon dioxide is via phos-phoenol-pyruvate (PEP) and is mediated by PEP-carboxylase which has a high affinity for carbon dioxide and thus is not rate limiting in the tropics.
Campos	Field in Portuguese and Spanish.
Catch Crop	A fast growing crop grown with or between the rows of a main crop.
Coevolution	Where species evolve with a change in an environmental factor of with another species.
Companion Planting	A method of growing plants together where plants provide cover, enhanced pest and disease resistance and/or nutrients for the companion crop.
Coombe	A valley closed at one end.
Dicotyledon	Plants with two seed reserves, cotyledons. Leaves have reticulate venation.
Dipterocarp	Trees that produce two winged seeds. Not all dipterocarps have two wings and three or five wings are not unknown.
Divided Slope Farming	Where the slope is divided into discrete sections for cultivation – aim is to reduce soil erosion.
Easement Programme	A conservation program in the United States where land can be set aside for conservation purpose.
Ecosphere	A closed, self sustaining ecosystem.
Endemism	Where species are unique to a given geographical area.
Entisols	Soils with no horizontal banding other than the A horizon.

Epiphyte	Free living plants growing on other plants, commonly on trees, and not parasitic.
Eutrophication	Nutrient enrichment of watercourses, ponds and lakes.
Fallow	To rest a field or loch (aquaculture) from continuous production.
Fly Strike	A term used to describe the laying of eggs into the fleece of sheep by blowfly
Forbs	Herbaceous plants in grassland.
Green Manure	A crop grown for incorporation into the soil, used in soil fertility building in organic systems.
Headage	Payments made on the basis of livestock numbers.
Heterogeneous Landscape	A rural landscape composed of many elements, hedges, rolling hills, watercourses, mixed farming, woodlands etc.
Horizon	A term used to describe distinct zonation in soil profile, such as the A horizon.
Low Input	A farming system designed to produce food products with minimal artificial inputs. Relies on low stocking density and crop diversity to achieve commercial success.
Maidens	A single stem tree, not a coppice or stool.
Mangal	A saline environment with mangrove and other tree species capable of life in saline soils.
Mange	A contagious skin disease caused by parasitic mites.
Meristem	Tissue in plants consisting of undifferentiated cells. Found in growing tips of plants
Mesic	A habitat that has a balanced supply of nutrients and moisture.
Monocotyledon	Plants where seedlings have one cotyledon. Veins in leaves typically run parallel with mid rib.
Monopodial	Trees that grow upward from a single growing point. Observed in understorey trees in tropical rainforest. Single stem.
Niche	A term used to describe the relationship between an organism or population in an ecosystem and with respect to each other.
Non-Point Source	A term used to describe environmental pollution that has no defined point of origin.
Poaching	A term used to describe the impact of cattle footprints on grass fields. Occurs in overstocked pasture, around gates and feeders and from keeping stock in pasture too long in the winter months.
Point Source	Used to describe a pollution event where the source has a defined origin. Example would be a leaking slurry store.

Policy Drivers	Policy that is developed to drive an industry down a desired route.
Policy Instruments	A term used to describe a set of policies that force industry to adopt a behavioural pattern. The aim may be the adoption of sustainable practice. An instrument may be a policy to support good practice.
Pollard	A method of managing the growth of a tree. Where the branches of trees are pruned from the main trunk at about 2.1m from the ground.
Prairie	Large expanse of natural grassland with herbs and shrubs. Moderate rainfall. Term refers to great plains of the United States.
Prescribed Grazing	Where a theoretical grazing regime is established and used to manage a habitat.
RAMSAR	International agreement on wetland conservation, RAMSAR Iran, 1971.
Ride	A woodland track, usually connects logging coupes to main extraction routes.
Riparian	Vegetation strip running parallel with river or stream system
Ruderal	Plants that have evolved to cope with high disturbance environments. Nutrients generally not limiting. Example is the arable field.
Savanna	Grassland with scattered trees, but no closed canopy. Ground cover is unbroken herbaceous layer of grasses and forbs.
Serpentine Soils	Soil derived from ultramafic rocks such as serpentine.
Soil Pan	A term used to describe compacted soil at the plough depth. Seen as a compacted horizontal band in a soil profile.
Standards	A term used to describe timber trees in a coppice rotation. Usually oak species that have been left for 8 to 12 coppice cycles.
Steppe	Grassland plain. Characterised by continental and semi-arid climate. Example the steppe's of Russia and highlands of Mongolia.
Strip Cropping	Where different crops are grown in strips across a broad and sloping/undulating landscape. The aim is to reduce soil erosion.
Sub-Strata	An underlying layer, such as the subsoil. Clay.
Supercooling	Where the aqueous environment of living tissue (plant and animal) can be frozen to a subzero temperature without freezing.

Swaling	A local name for burning of moorland habitats.
Sympodial	Where the crown of the tree is spreading and broad. Observed in emergent trees and isolated trees in hedges and fields.
Tiller	A side shoot in grasses and cereals.
Vernalisation	To make spring like. Plants need to experience a prolonged period of cold $<8°$ C to induce flowering.
Volatile Organic Compounds	Organic chemical compounds that vaporise at room temperature. There are many naturally occurring volatile organic compounds but there are also numerous harmful or toxic volatile organic compounds derived from modern industry.
Xeric	Sun baked and dry habitats

Index